A Day at the Fair

SKETCHES FROM
THE ROYAL AGRICULTURAL
WINTER FAIR

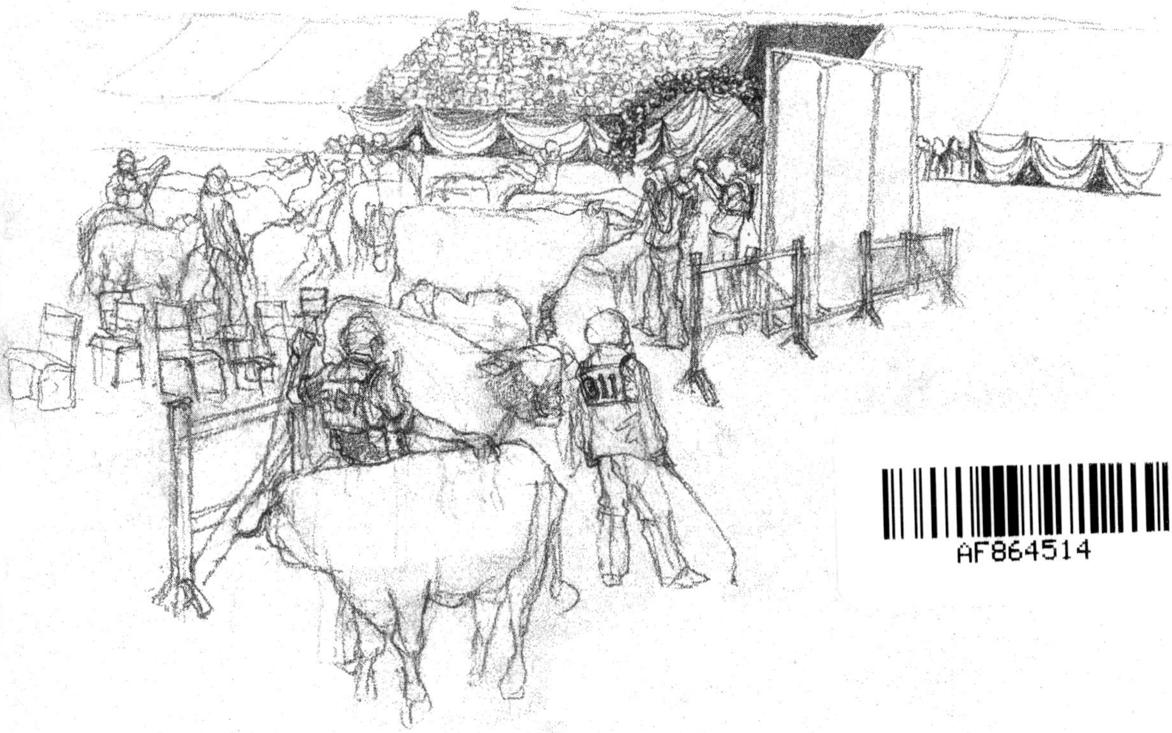

Sketches and line drawings by Paul Young
With a historical text by Paul King

KEY PORTER BOOKS

Canadian Cataloguing in Publication Data
Young, Paul, 1937–
A Day at the Fair: sketches from the Royal Agricultural Winter Fair

ISBN 1-55013-677-1

1. Royal Agricultural Winter Fair (Toronto, Ont.) – Pictorial works. 2. Royal Agricultural Winter Fair (Toronto, Ont.) – History. I. King. Paul, 1935–
II. Title.

S557.C32T67 1995   630.74713541   C95–931754–6

Key Porter Books Limited
70 The Esplanade
Toronto, Ontario
Canada M5E 1R2

The publisher gratefully acknowledges the assistance of the Canada Council and the Ontario Arts Council.

Design: Peter Maher
Printed and bound in Canada

95 96 97 98 99 6 5 4 3 2 1

Contents

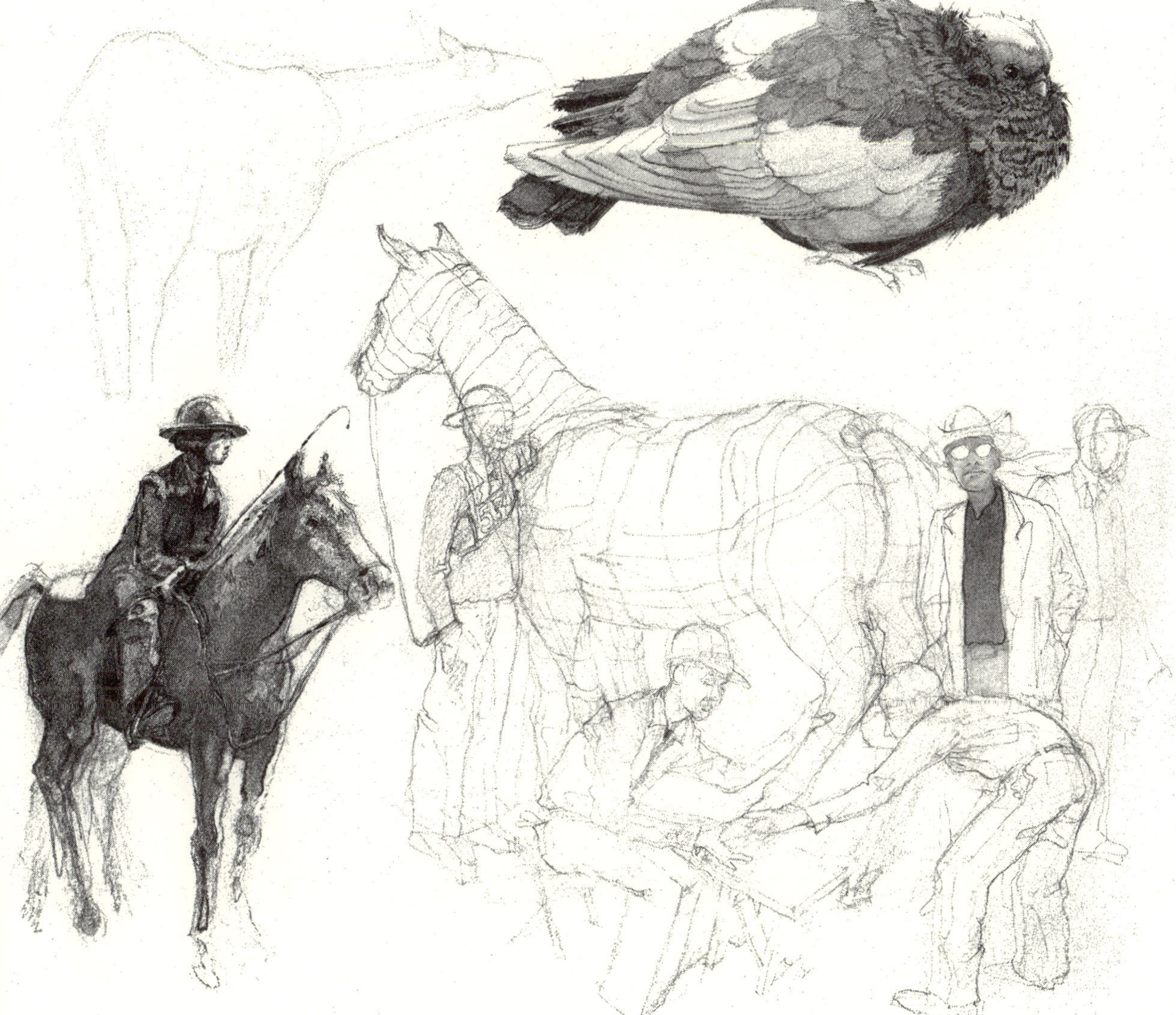

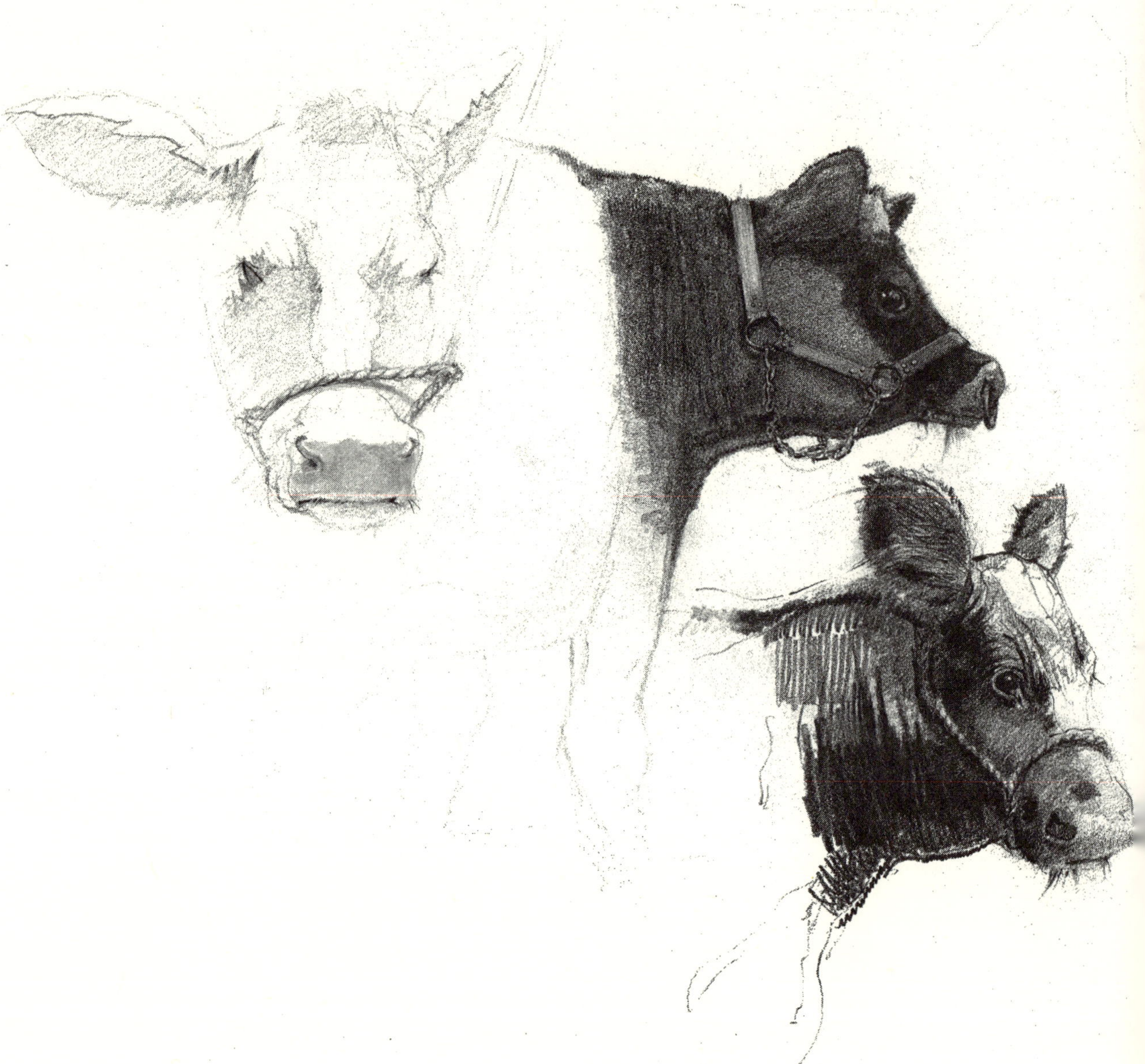

Preface by the Artist

A SKETCHBOOK IS NOT, SUI GENERIS, A WORK OF art. In most cases it is merely a vessel, a visual journal that unravels or fleshes out some inquiry. It may be central to an artist's path or entirely tangential. The sketches reproduced within these pages span more than twenty years of visits to the Royal Agricultural Winter Fair. Accordingly, the subject matter is highly focused, even if individual elements vary in style and intention. Everybody goes to the Royal with different motives, and in a Tolstoyan sense, each takes away something unique from the experience. It is not difficult to be impressed by a team of spangled Clydesdales. But for some, there is parallel majesty in a fine sheaf of winter wheat or the largest pumpkin in the province.

Like most native Torontonians, I began attending the Royal as a child. Usually it was a family outing, occasionally a school trip. One always noticed the large number of artists scattered about, an early indication that something very important was happening. I later learned that they were in fact art students, detailed to pit their youthful skills in observational drawing against most difficult subjects under abnormally trying circumstances. The majority probably came from the Ontario College of Art, but other worthy schools such as Central Tech used the vast resource of the Royal to enhance their art programs. During my own student years at OCA, it was the habit of the drawing and painting department to virtually shut down normal operation throughout the all-too-brief run of the fair. Later the faculty would jury and hang a large in-house exhibition of sketches. Inclusion was considered a signal honour, but the greatest coup of all was to be appointed that zaniest of zanies, a butter sculptor. Happily, student carvers continue to work their yellow masterpieces within the chilly glass-walled "studios" adjacent to the Coliseum, a fine and preposterous tradition much appreciated by veteran visitors.

For a neophyte artist, the fair is a supreme test of concentration. Working in public is always a bit nerve-racking, but drawing at the Royal is something else again. At all times the noise is shattering. In all places swarms of happy people jostle about and are to be forgiven if they sometimes think the sketchers are additional public exhibits to be prodded, questioned and logged into the class daybook. At the very centre of these distractions are the animals themselves, who simply refuse to stay still. Like disobedient Olympias they defy the artist, lull him into thinking this or that position will be sustained, only to dash his hopes by moving to another part of the forest or doing something distasteful. Nubian goats form demonic conspiracies, for while one long-eared odalisque pretends to pose, another blindsides the artist and chews his sketchbook with relish. It is both maddening and hilarious.

There is a profound lesson concealed within the hubbub. Only the most gifted students learn it quickly. To draw an animal, it is first necessary to understand it. Constantly shifting contours challenge the innocent copycat to a battle of wits, and without a firm grasp of how a creature is actually built, its edges remain forever elusive. The entire enterprise devolves upon the recognition of architecture and solid geometry in motion. It is prudent to study the subject very carefully, simplifying the masses and planes, rehearsing and memorizing an entire repertoire of postures. For, sooner or later, every beast returns to the situation in which it was first discovered. Consequently, artists spend more time looking than drawing. They are hunters, not spectators.

It may seem that the purpose of the exercise is to make art, but such an exalted conceit hardly applies to the many thousands of students who have been tested at the Royal in this century. For them, paper and pencil are not quite a medium, but primary tools of analysis and revelation. The sketch that results may be attractive in its humble way, but more retrievable value lies in the drawing process itself. Each picture is just a step in the ascent, a talisman of a lifelong odyssey. In our world of deconstructed post-modern folly, we must remind ourselves that an artist who cannot draw is no artist at all. That is why we go to the Royal, to put ourselves at risk. The animals help us tough it out because they are so interesting, so alluring.

At first glance, there seems to be some gulf between the country-bred exhibitors and the largely city-bred artists. The former obviously have some genuine affection for their beasts and think of them as more than money on the hoof. But always they must deal with the facts and the familiar in husbandry. Perhaps they smile in the knowledge that for us city mice, the prize-winning Holstein bull may be the most exotic and terrifying thing we see that year, possibly in a lifetime. As to the judging of animals, this is central to the exhibitor's purpose. But to the urban observer it seems to be a mysterious beauty pageant, some esoteric ritual trial where all the acolytes are robed in white. The presiding judge reviews the evidence and intones the verdict, but usually in a language spoken only by adepts. The animals seem not to care if they win or lose, if their next destination is the Elysium of breeding or the abattoir.

Forty years have elapsed since my own college days. Now I am the drawing teacher, and it is my students who must acquit themselves within the maelstrom. I have never missed a fair and long ago built special sketchbooks to the purpose. The sketches in this book are excerpted from the original books. They are just anecdotes. There is no room for art at the Royal; it is enough to survive.

Every fall I take the Red Rocket down Bathurst clutching my book, my stool and my pencils. Gradually a familiar calm invades the spirit because soon something astonishing will appear, a vision so wonderful that it will command every scrap of experience and interior focus. It may seem paradoxical that in the middle of calamitous noise one can be perfectly serene, but art is full of paradoxes and this special one is the most enduring of all.

I speak for the legions of artists in expressing my gratitude to the Royal's directors. The complex event they orchestrate each November is of the utmost importance. The accommodation and encouragement they have historically extended to the art community, in particular, is beyond measurable value. Thanks must also be given to the exhibitors. Perhaps the gulf between is not so large, after all. Everything they produce has meaning, from the smallest piglet to the most soldierly line of rutabagas. All of it is beautiful and bears the imprint of that environment wherein they are the principal custodians. Although no amount of drawing has ever instructed me in the singular toils of agriculture, it has not left me blind to the wonder of creation. For this I thank my drawing masters, Eric Friefeld and Frederick Hagen, to whom this book is dedicated.

Plates

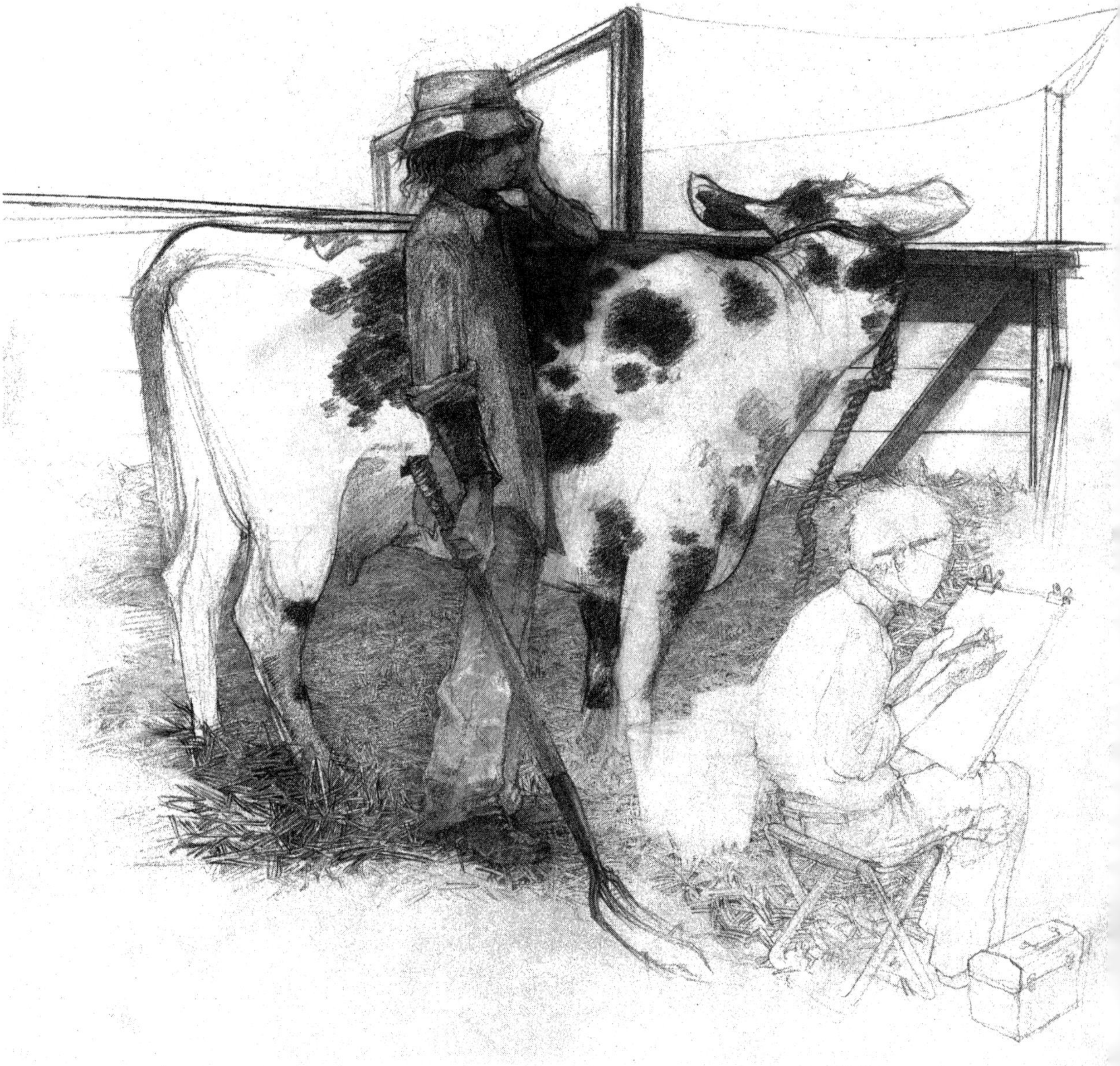

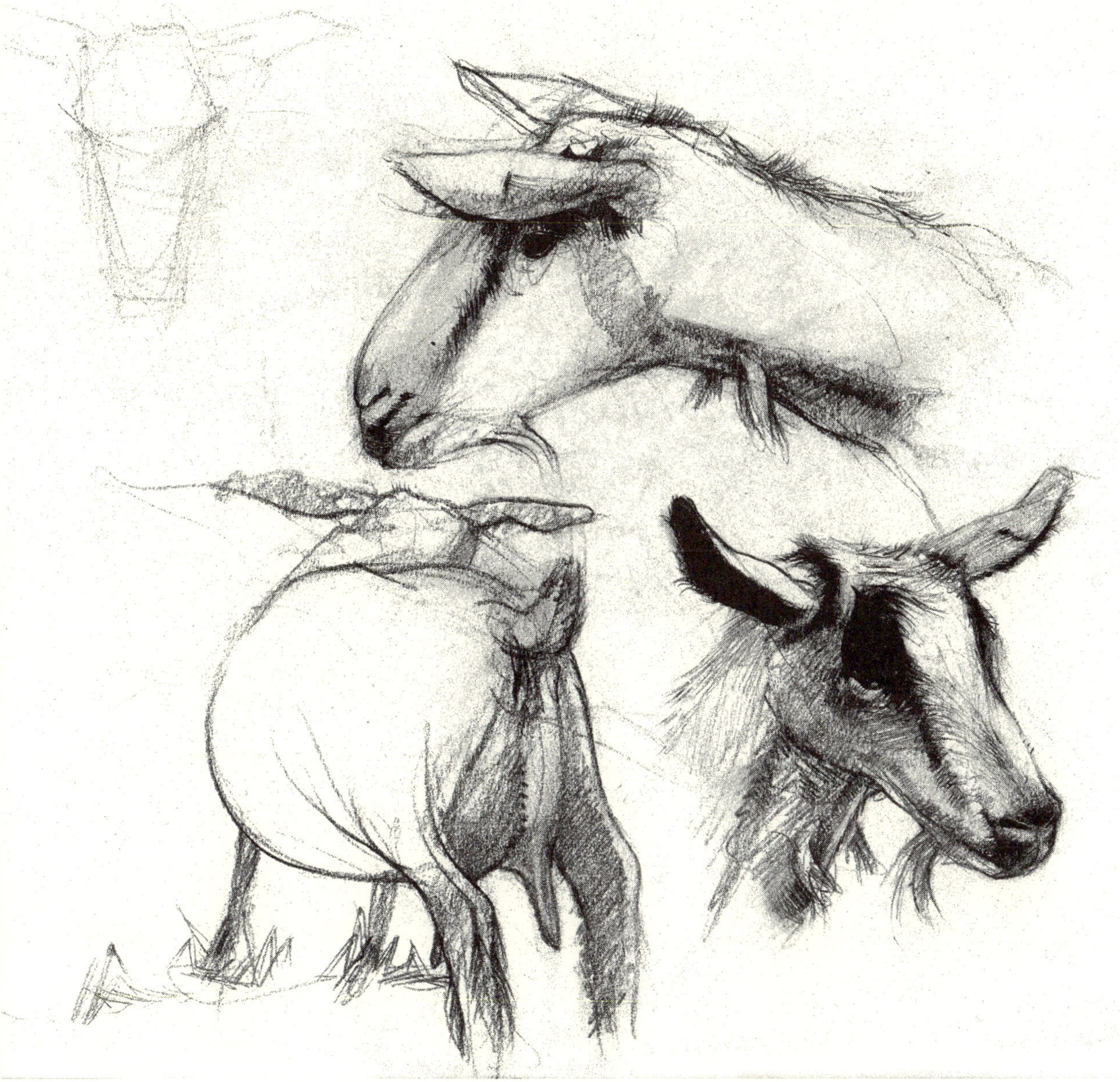

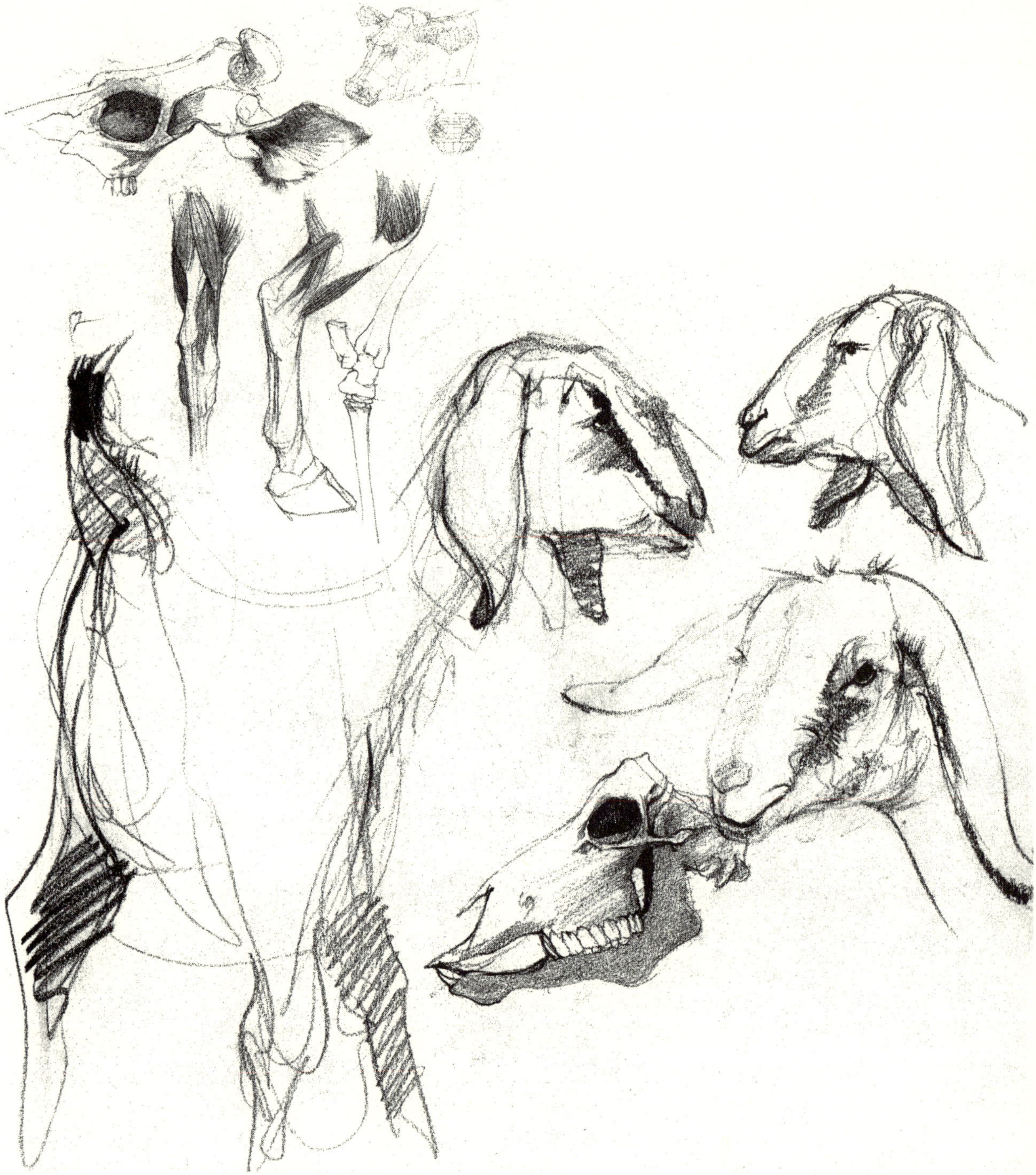

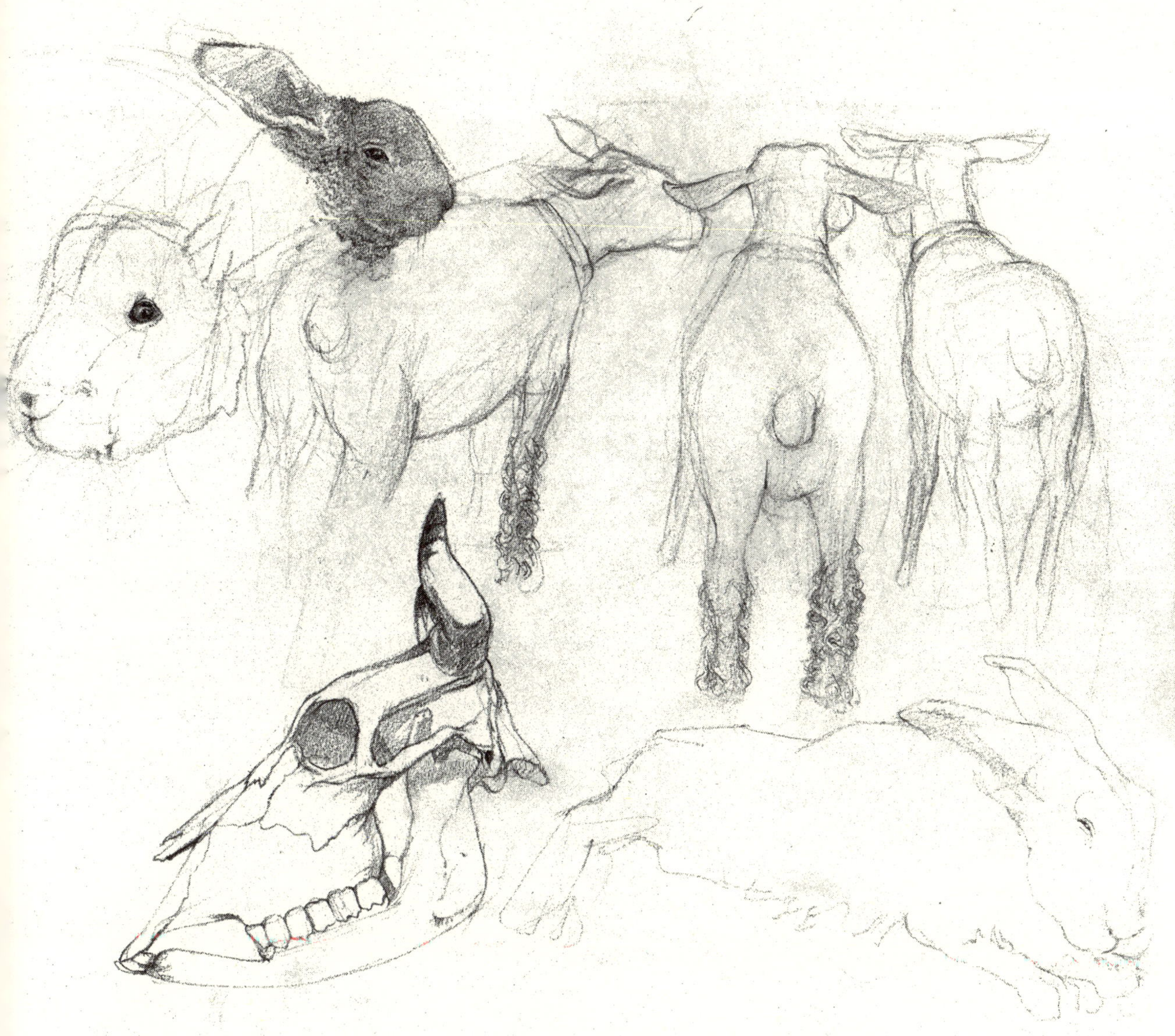

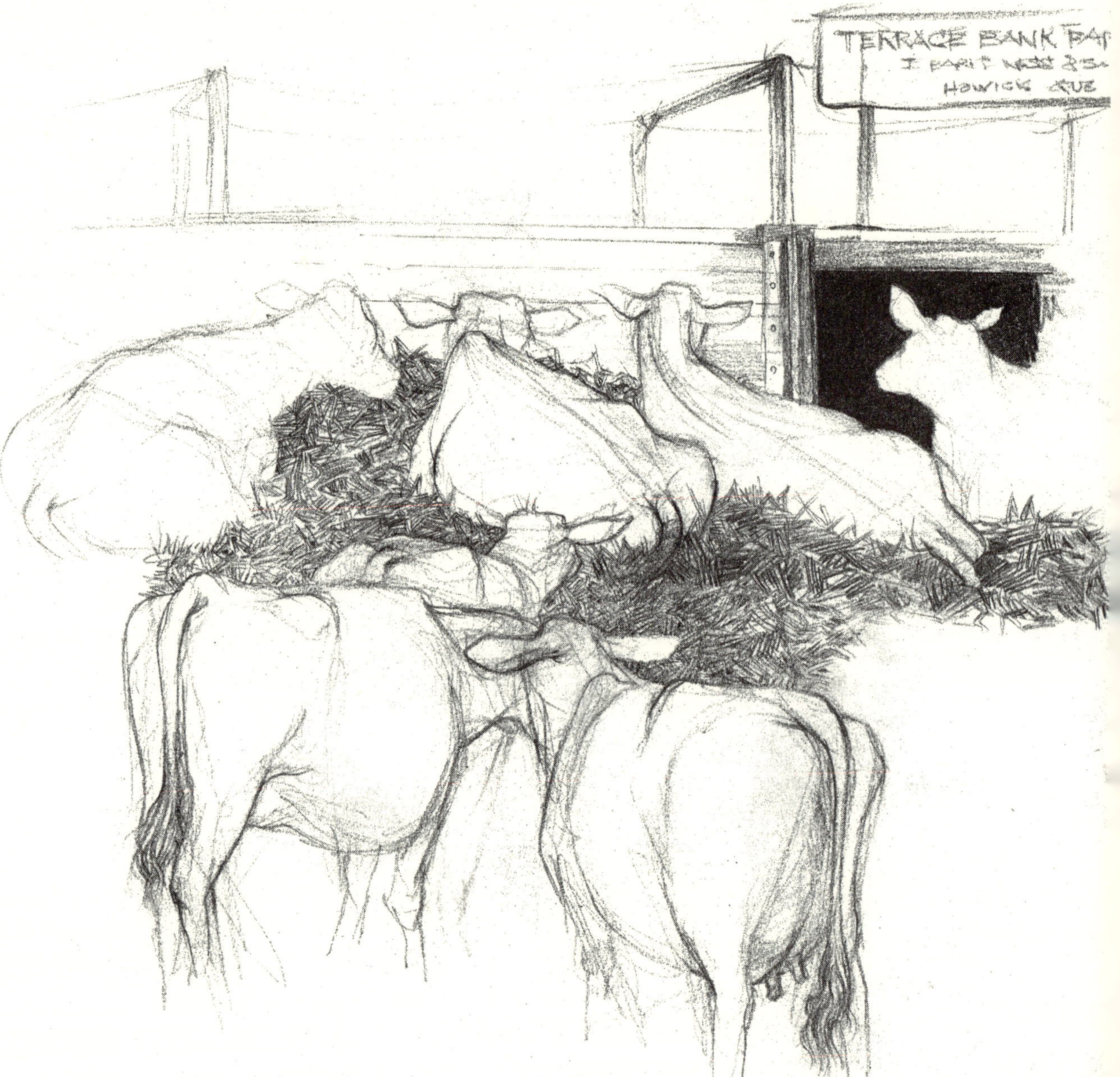
TERRACE BANK
HOWICK

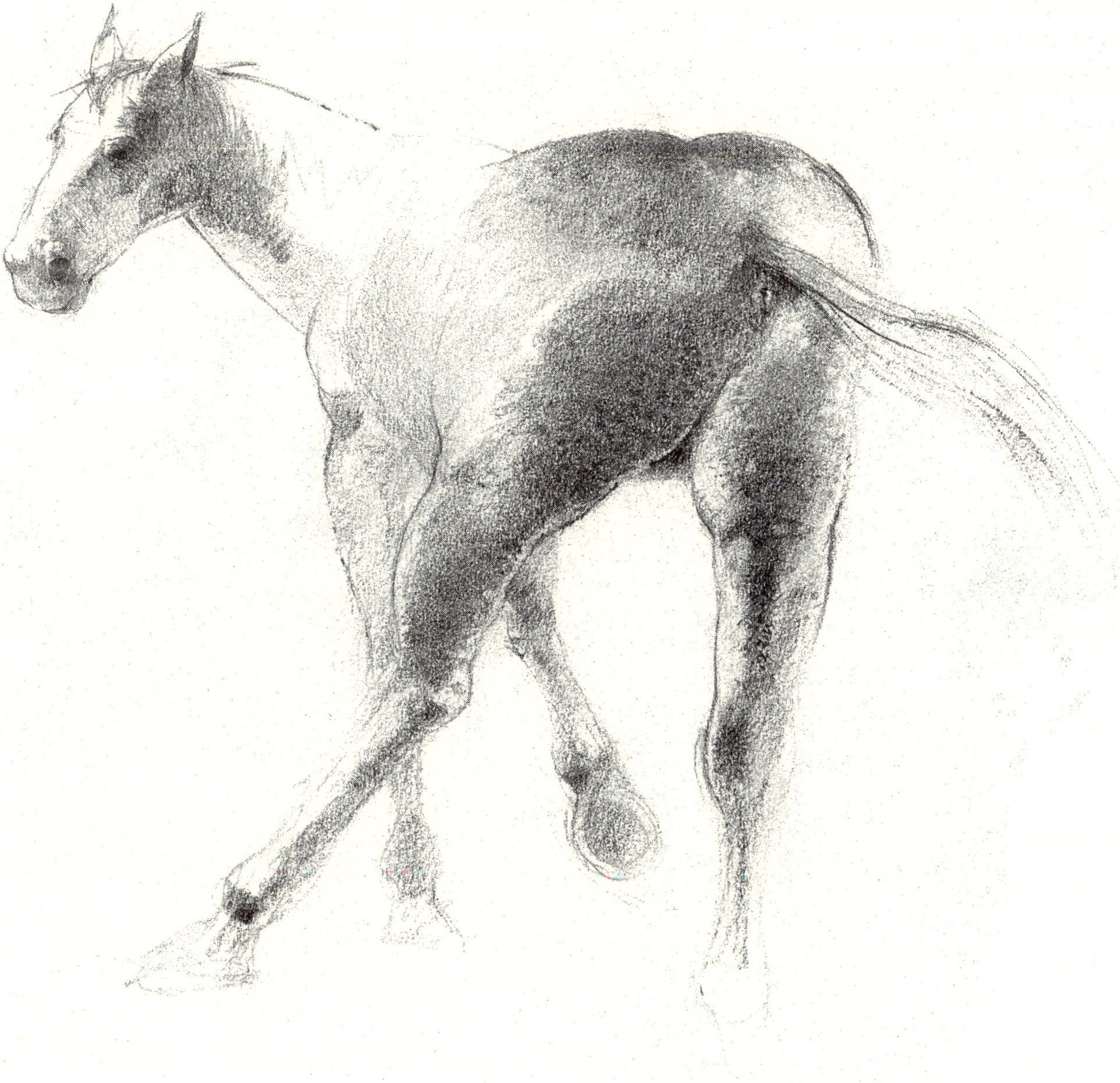

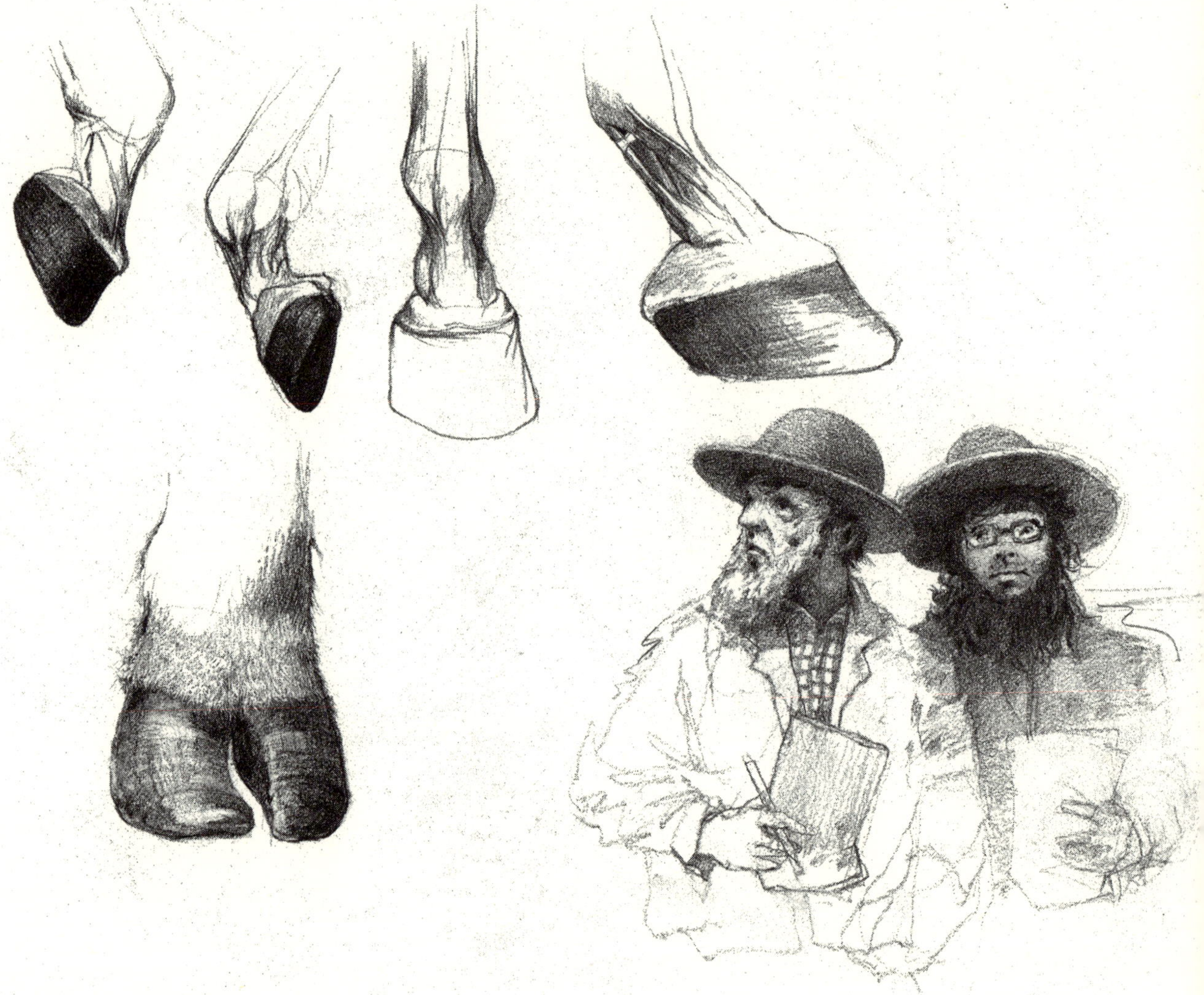

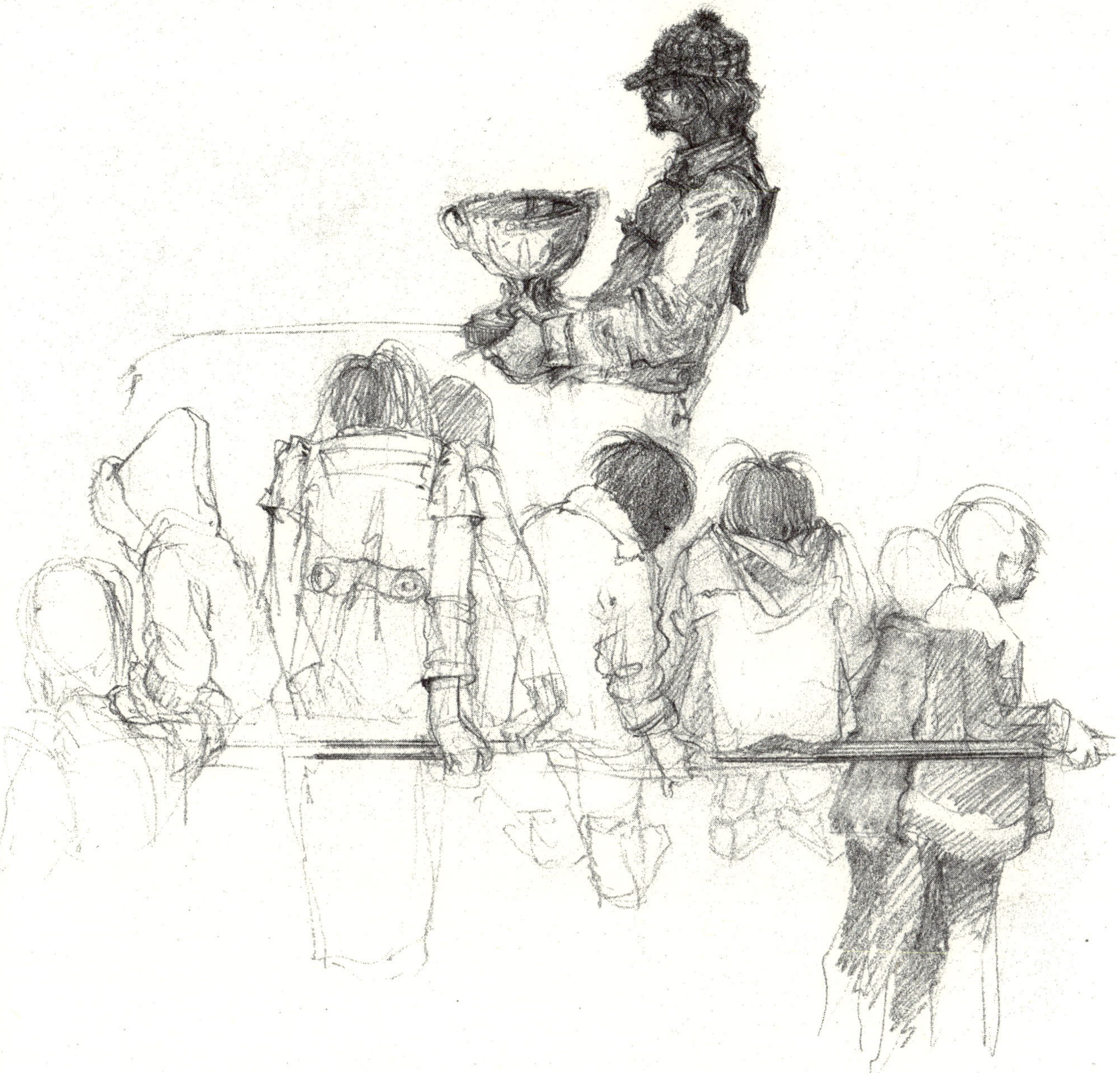

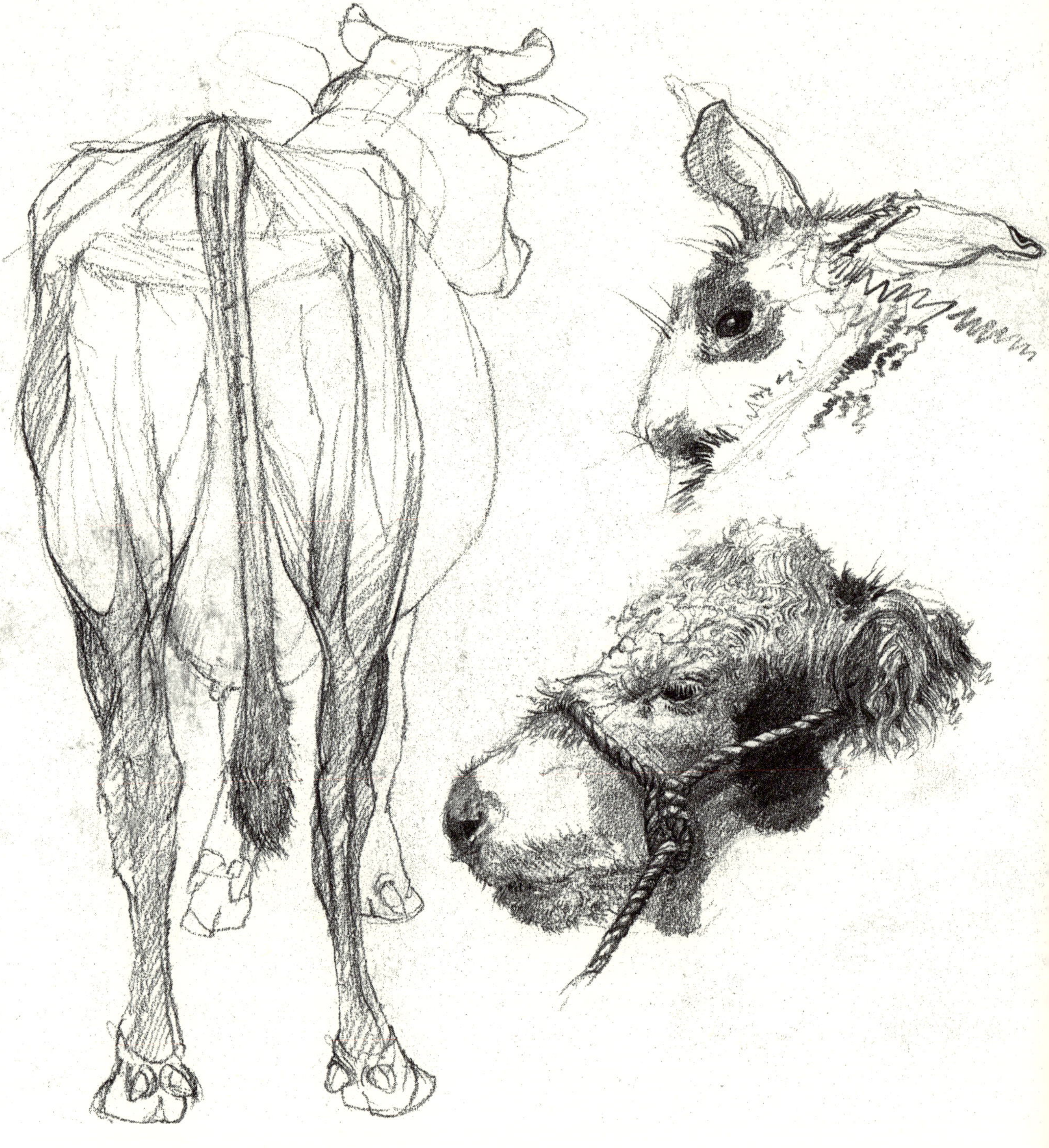

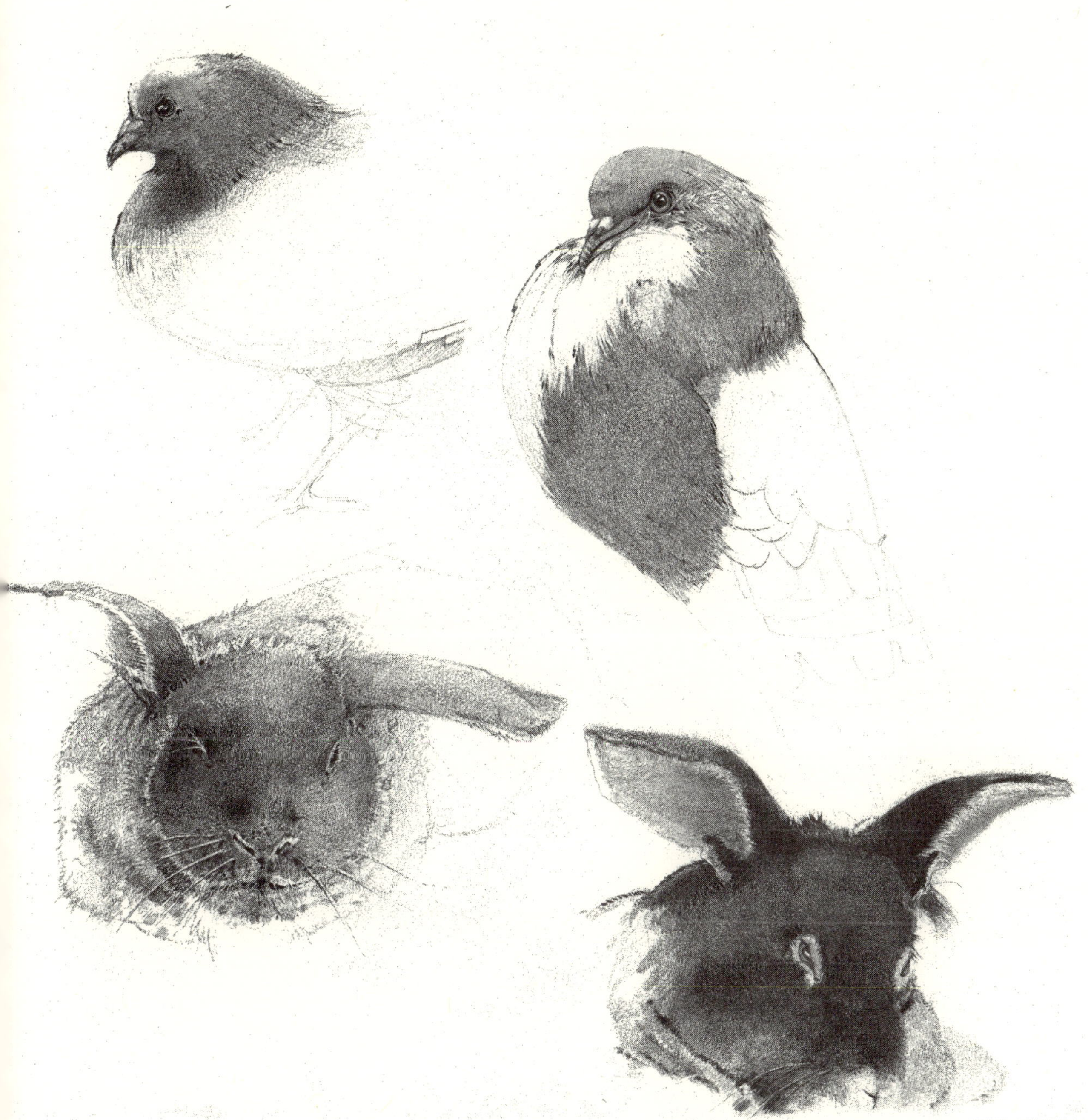

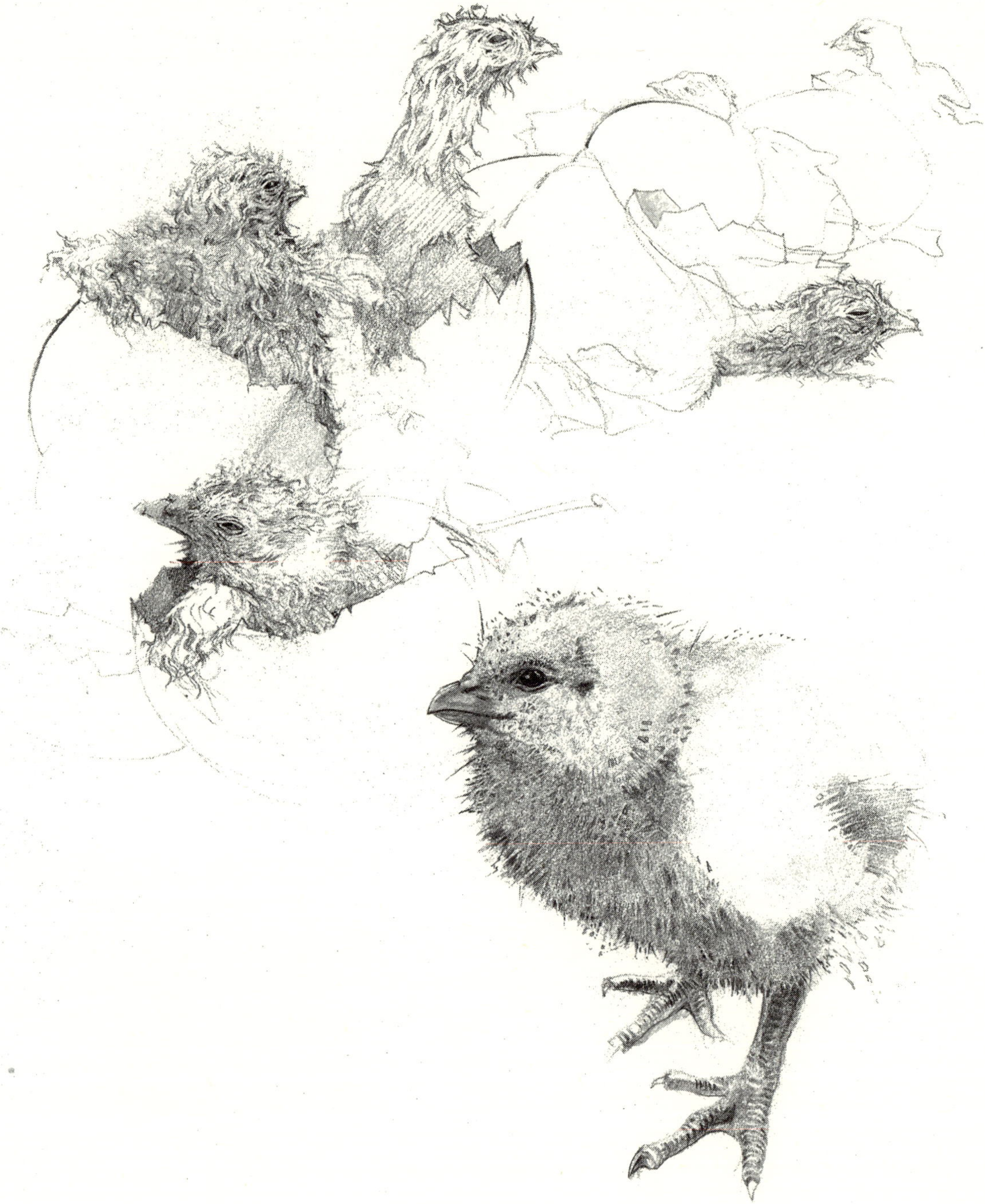

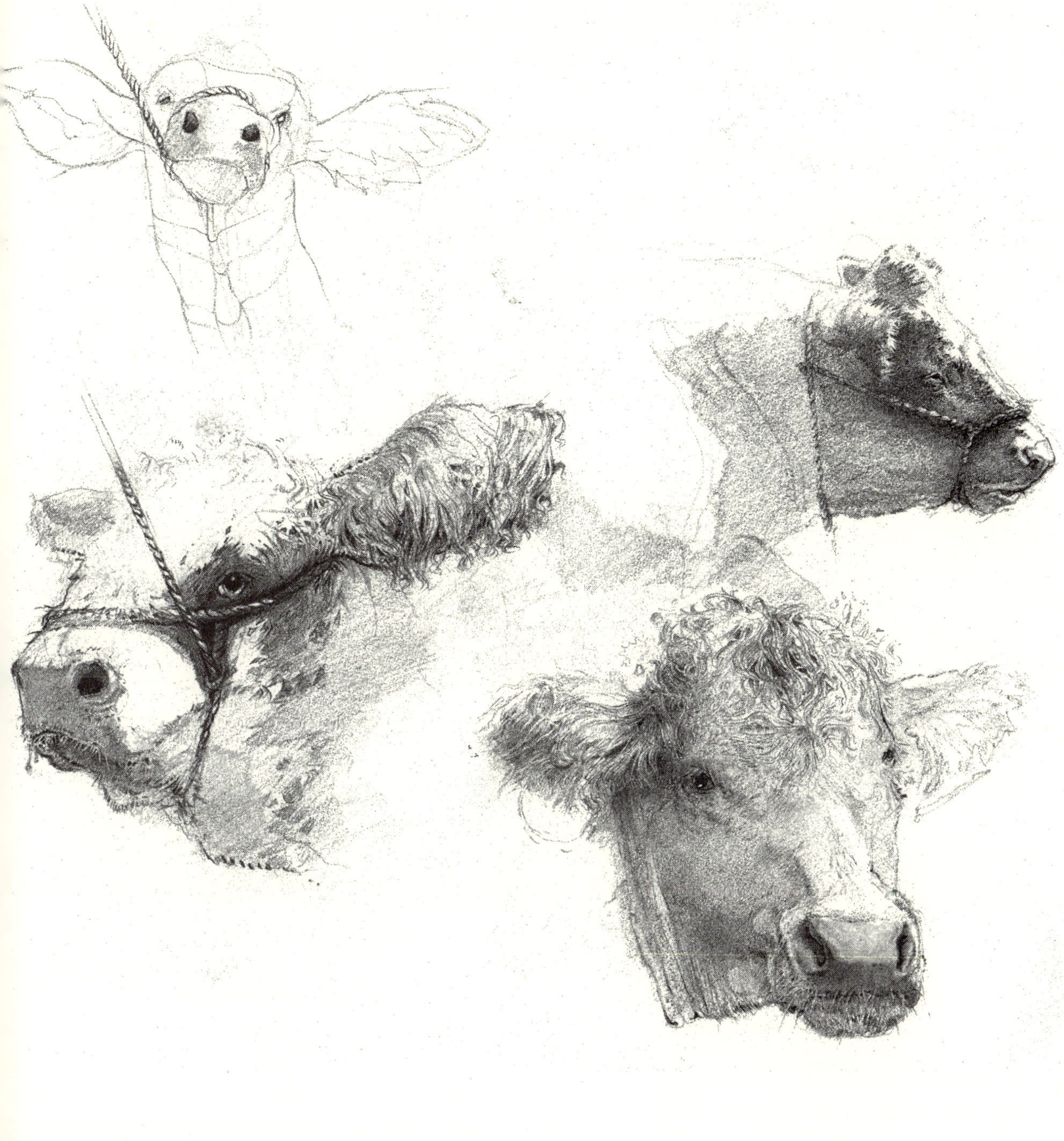

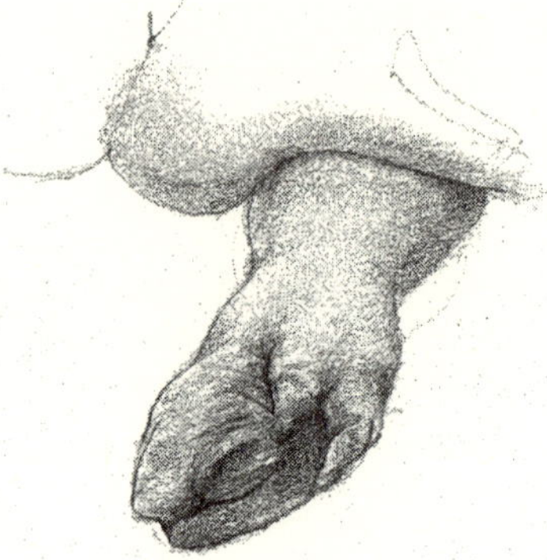

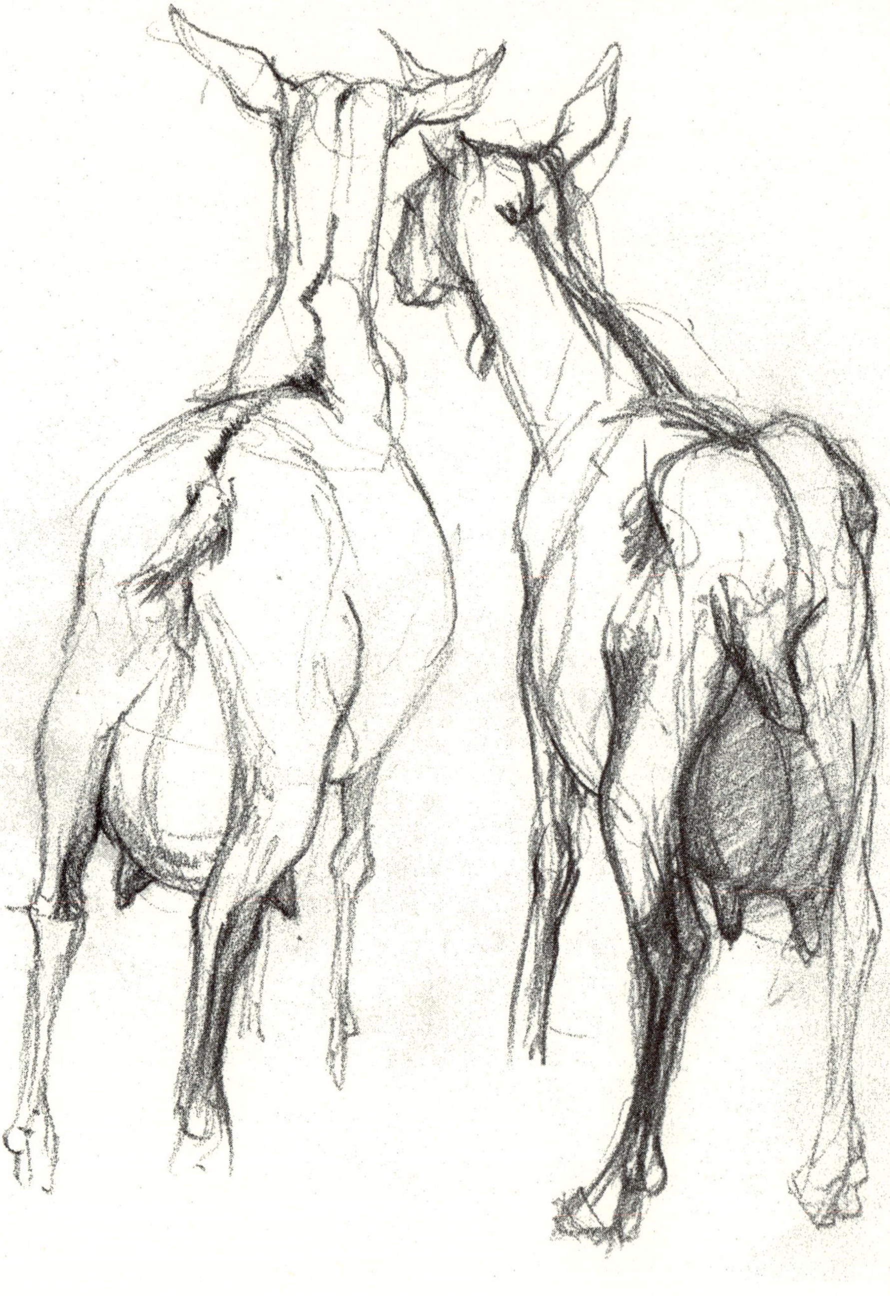

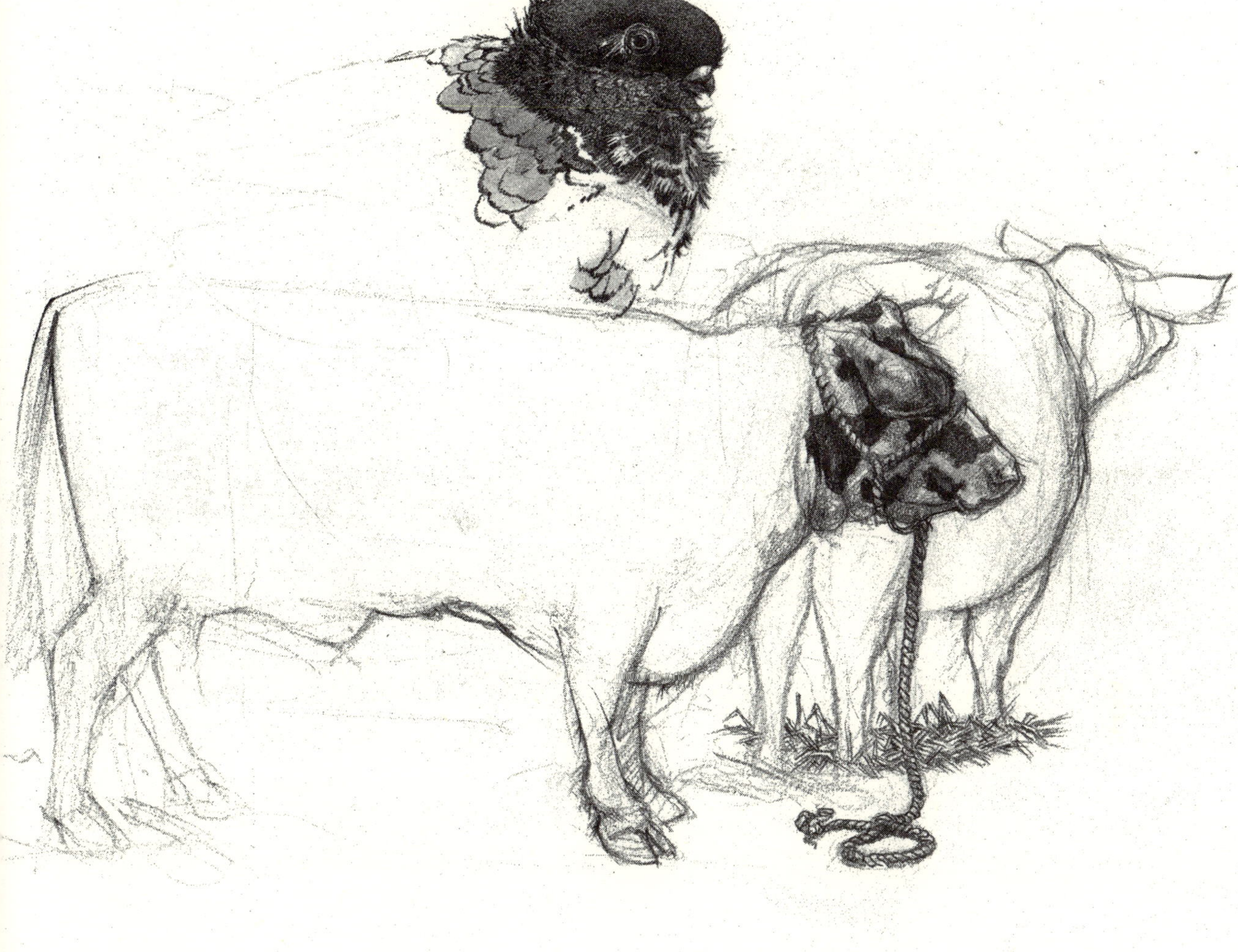

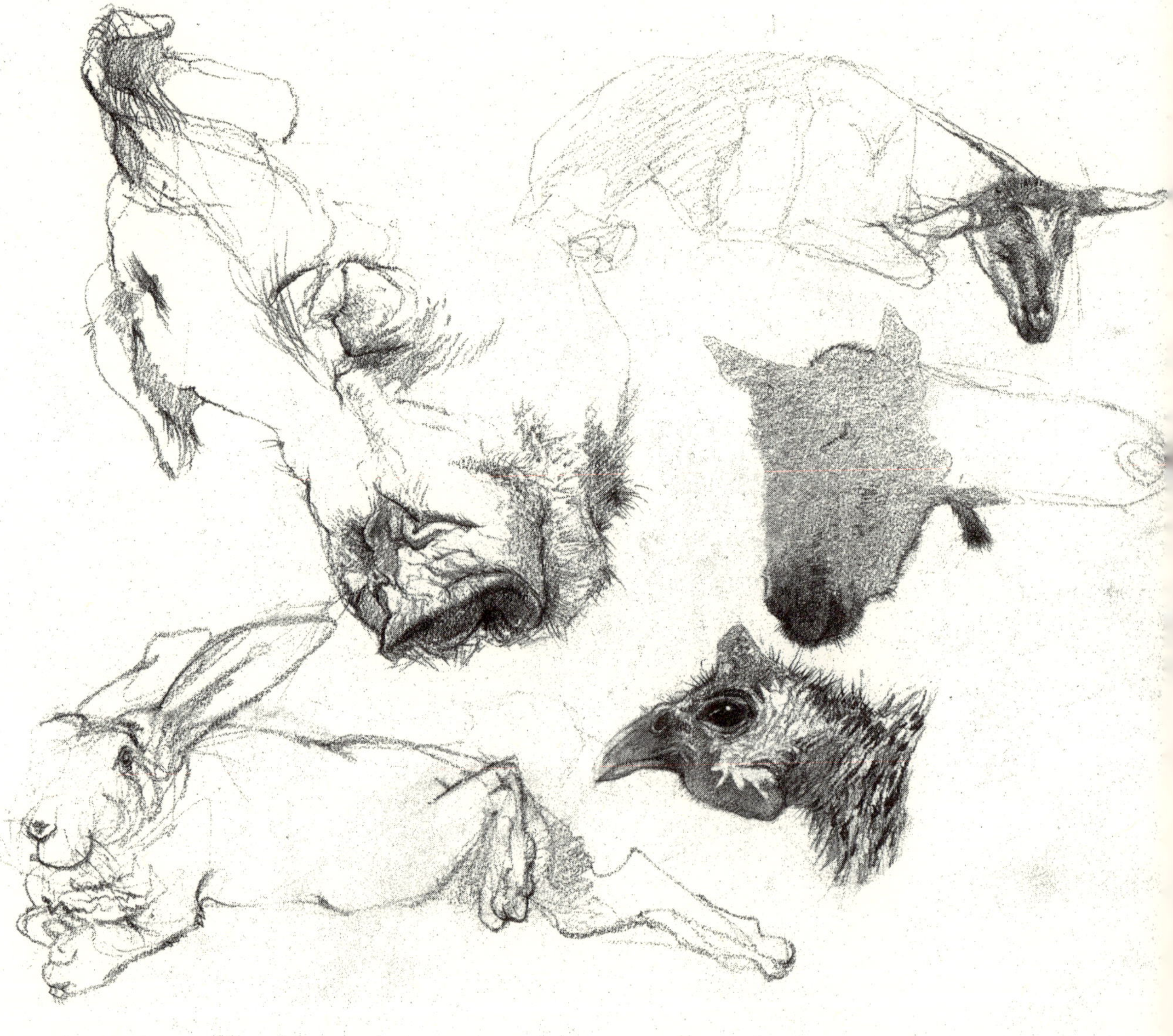

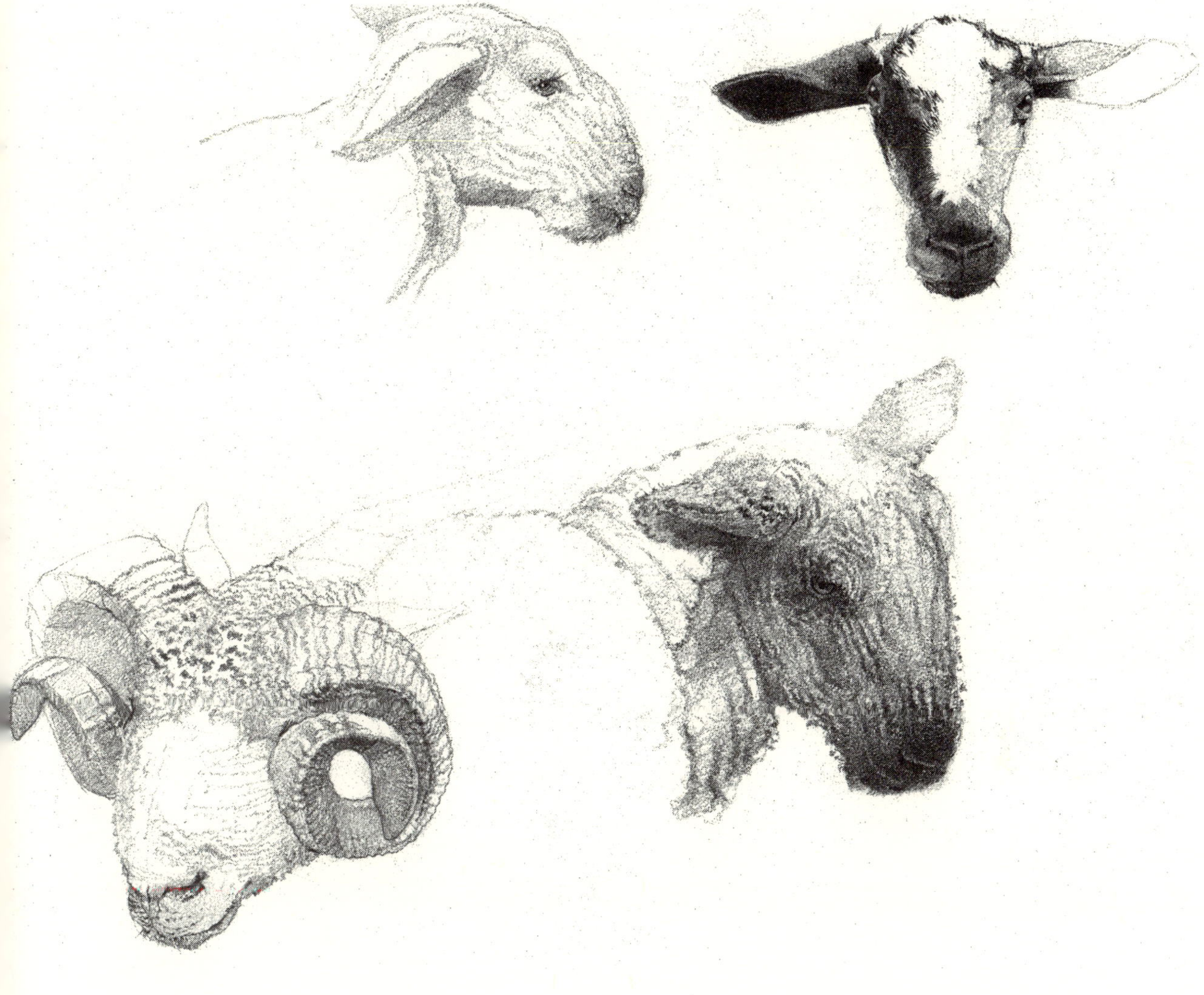

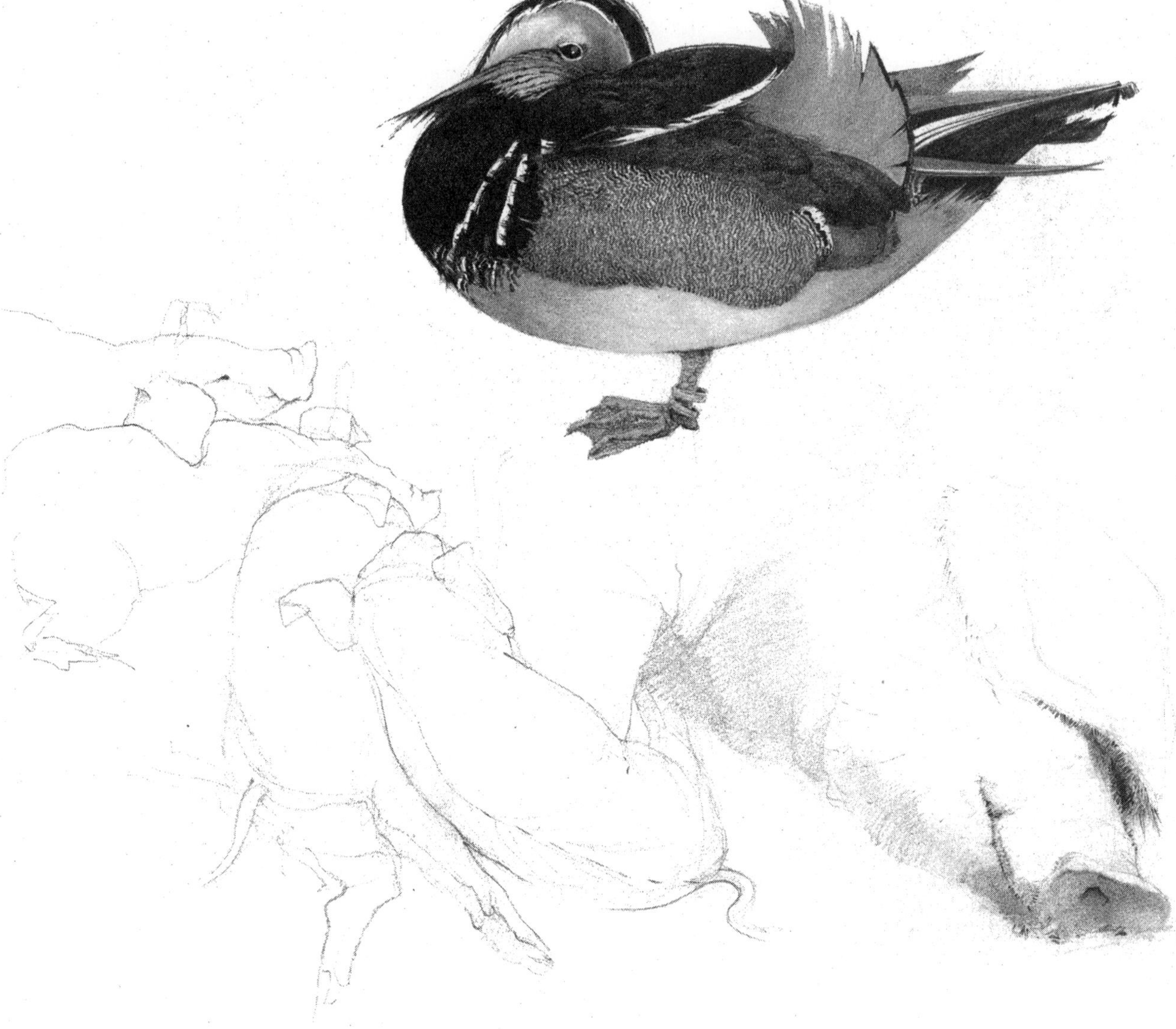

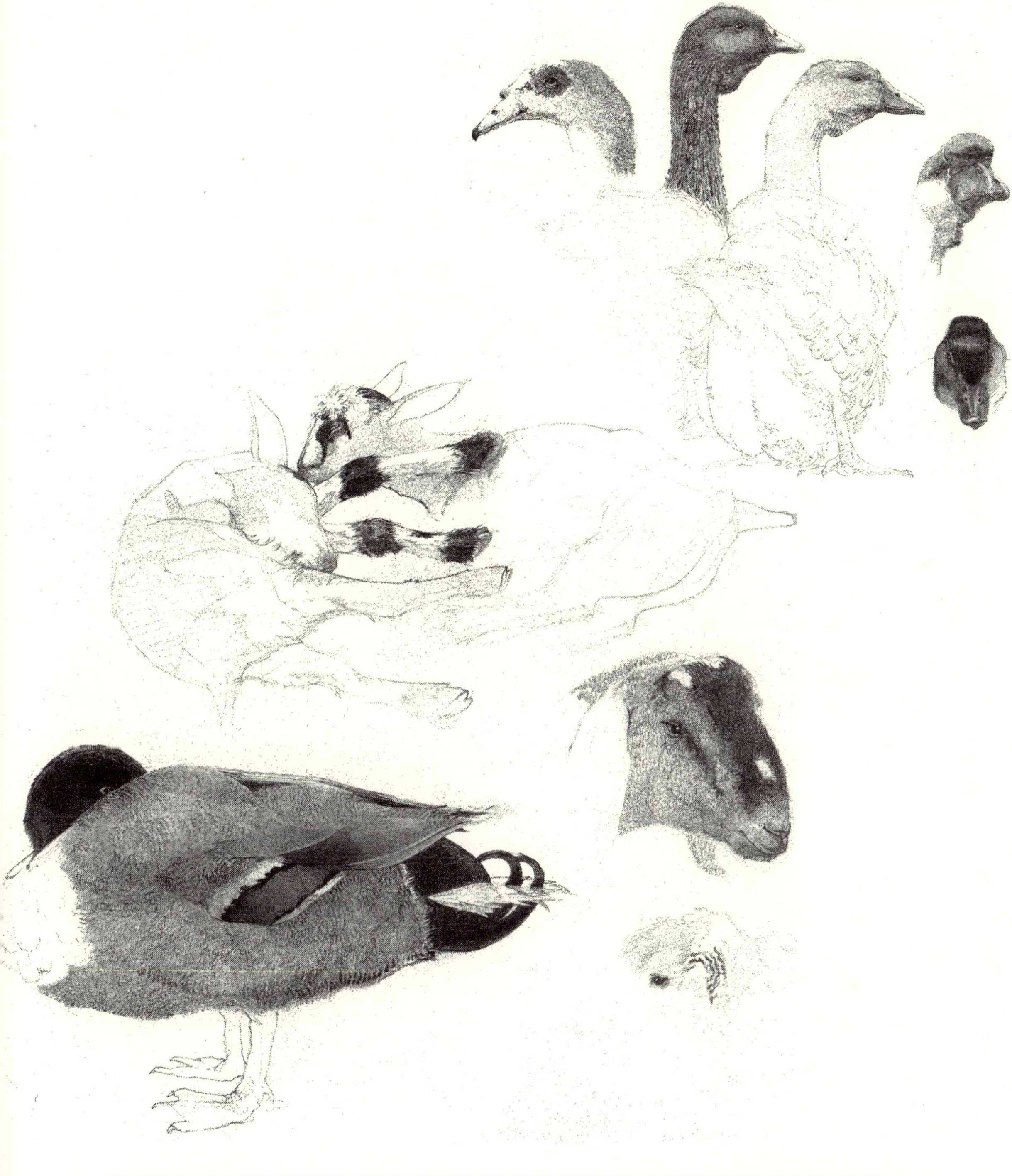

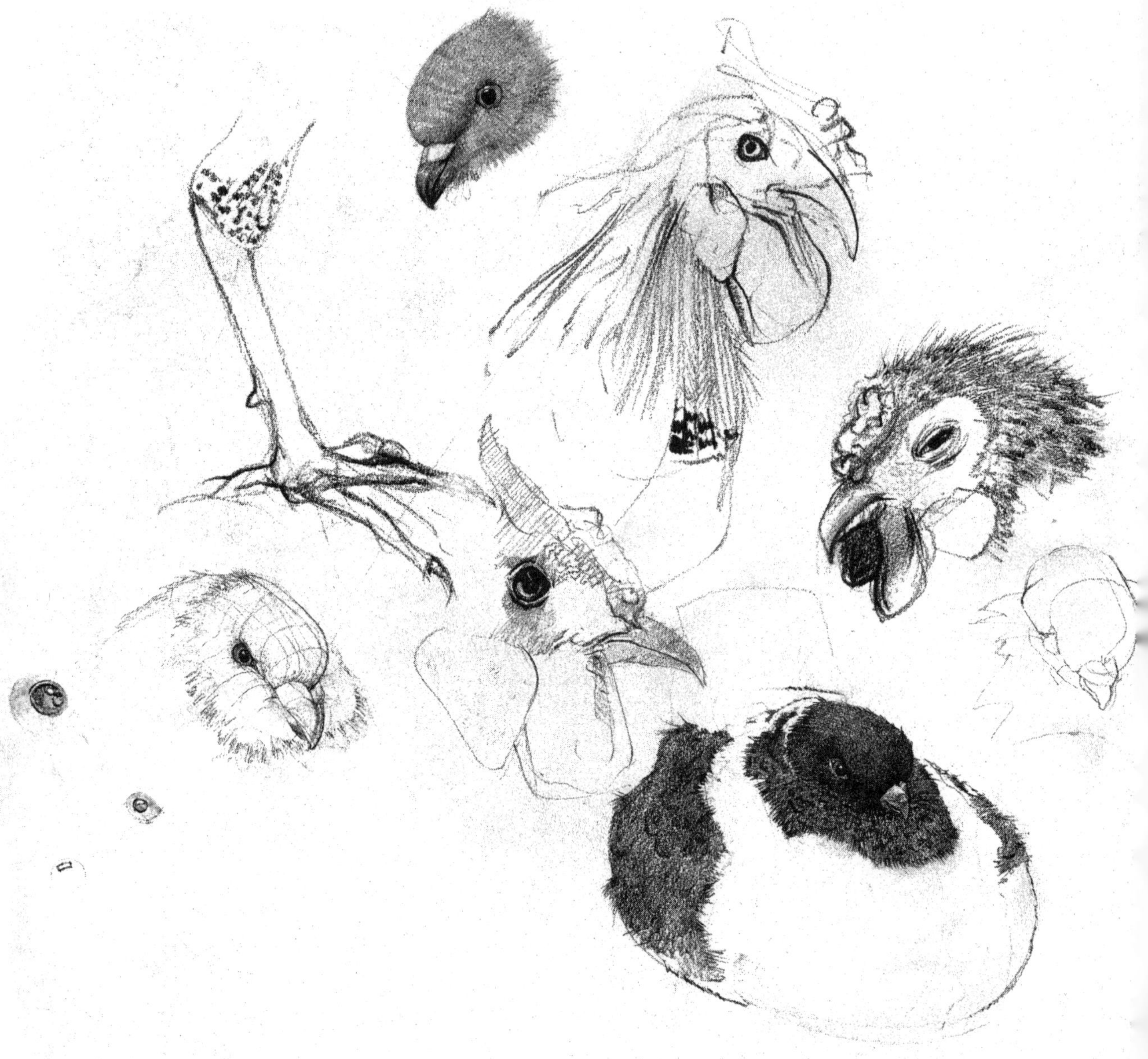

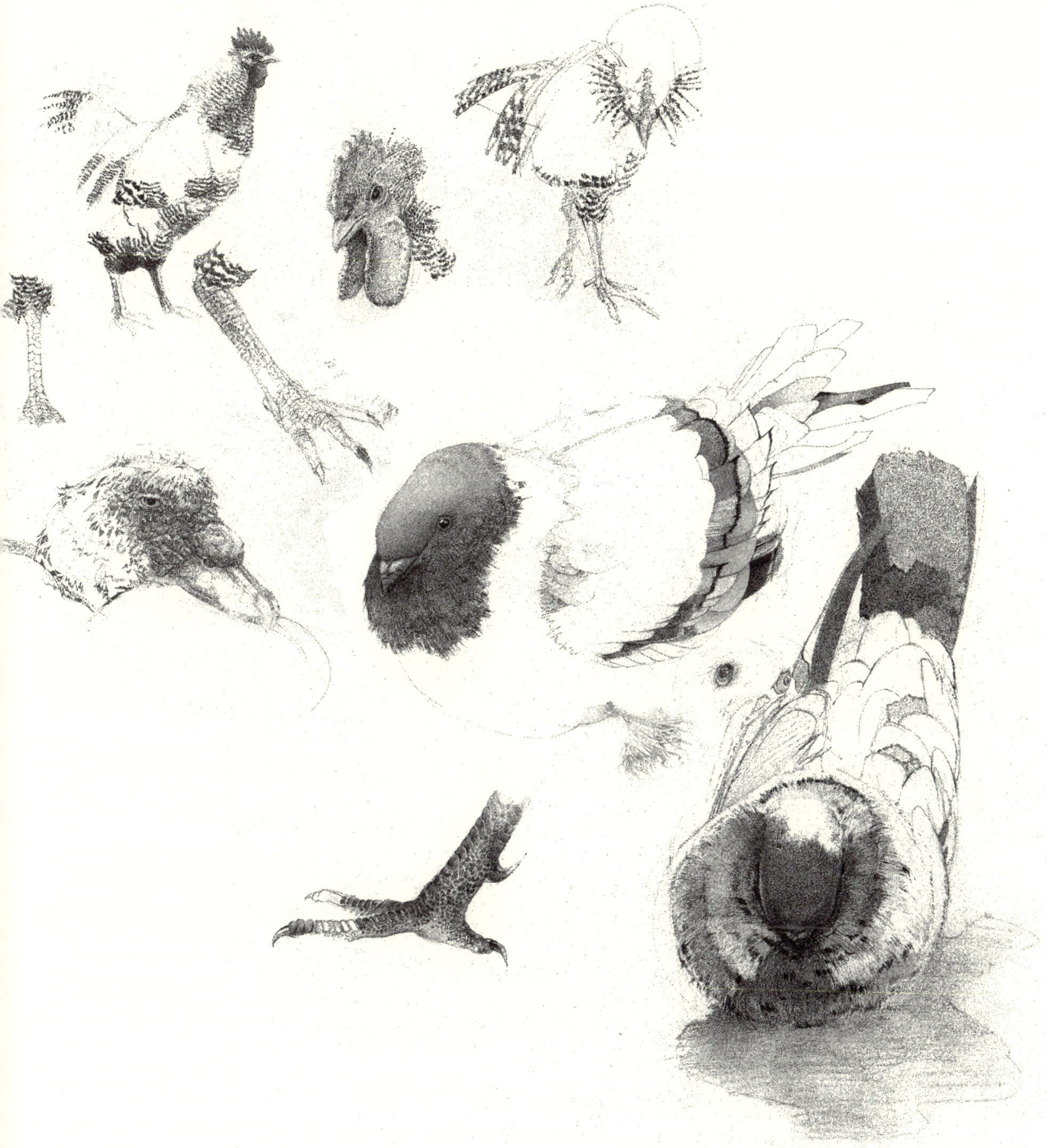

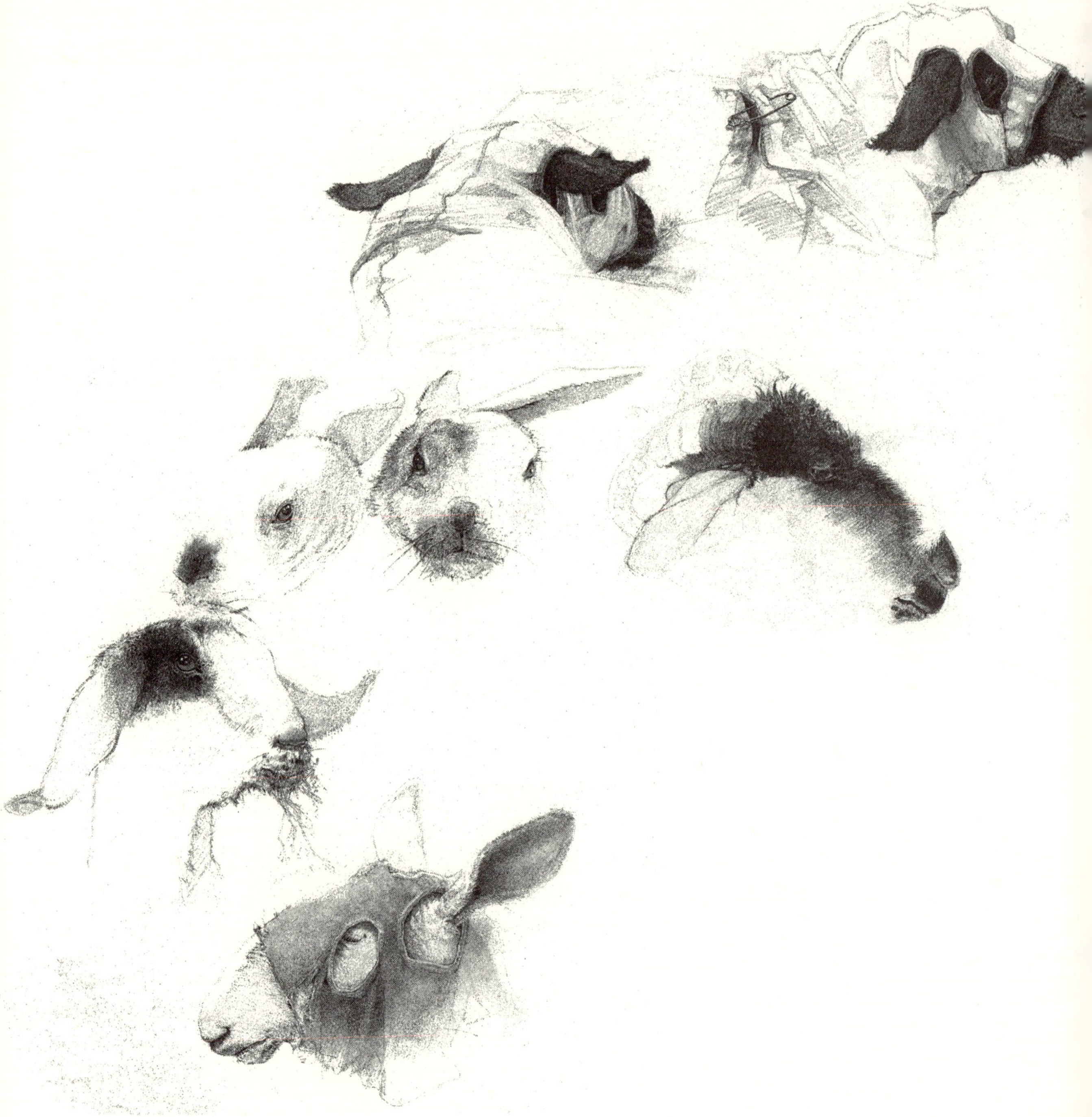

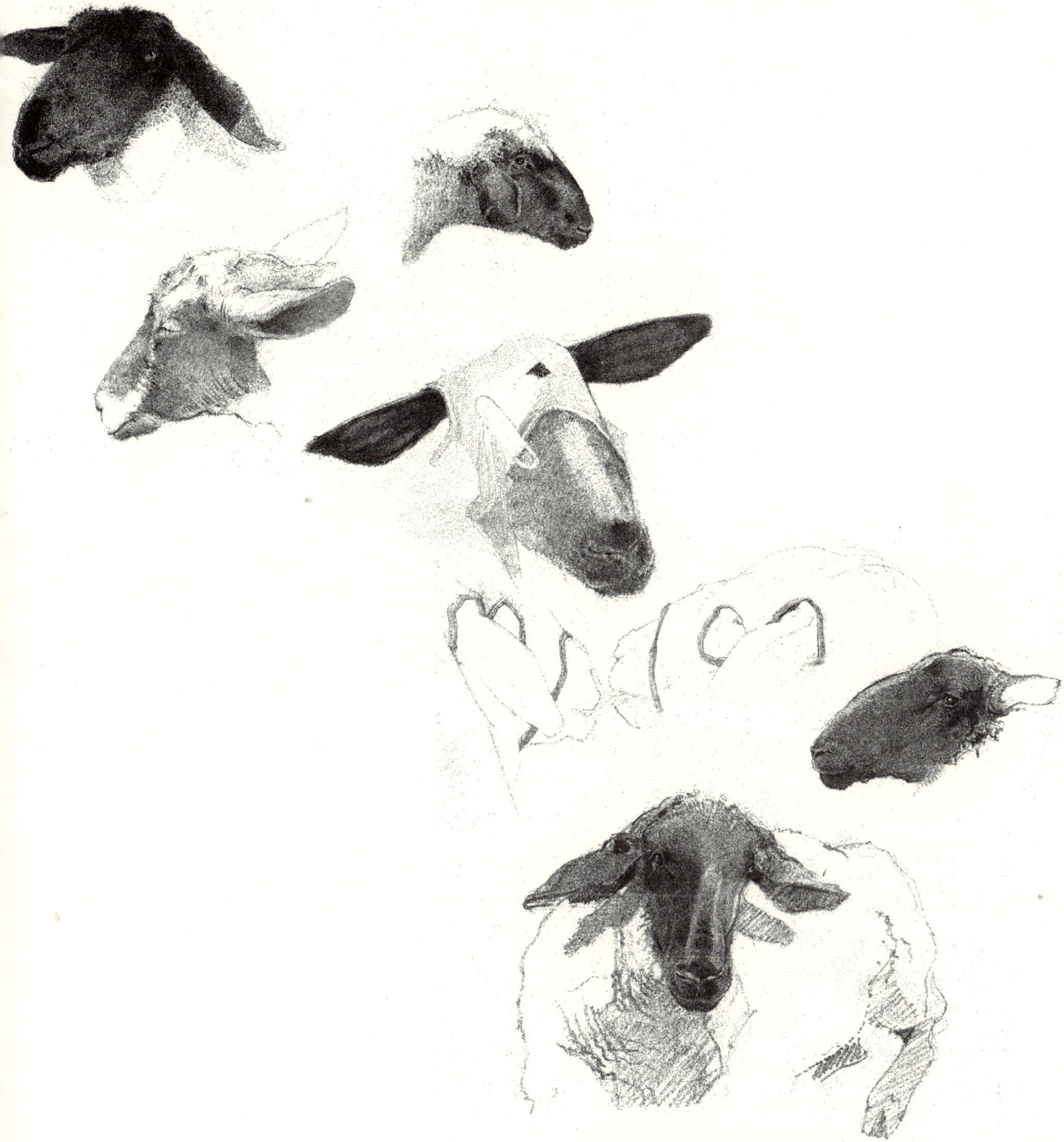

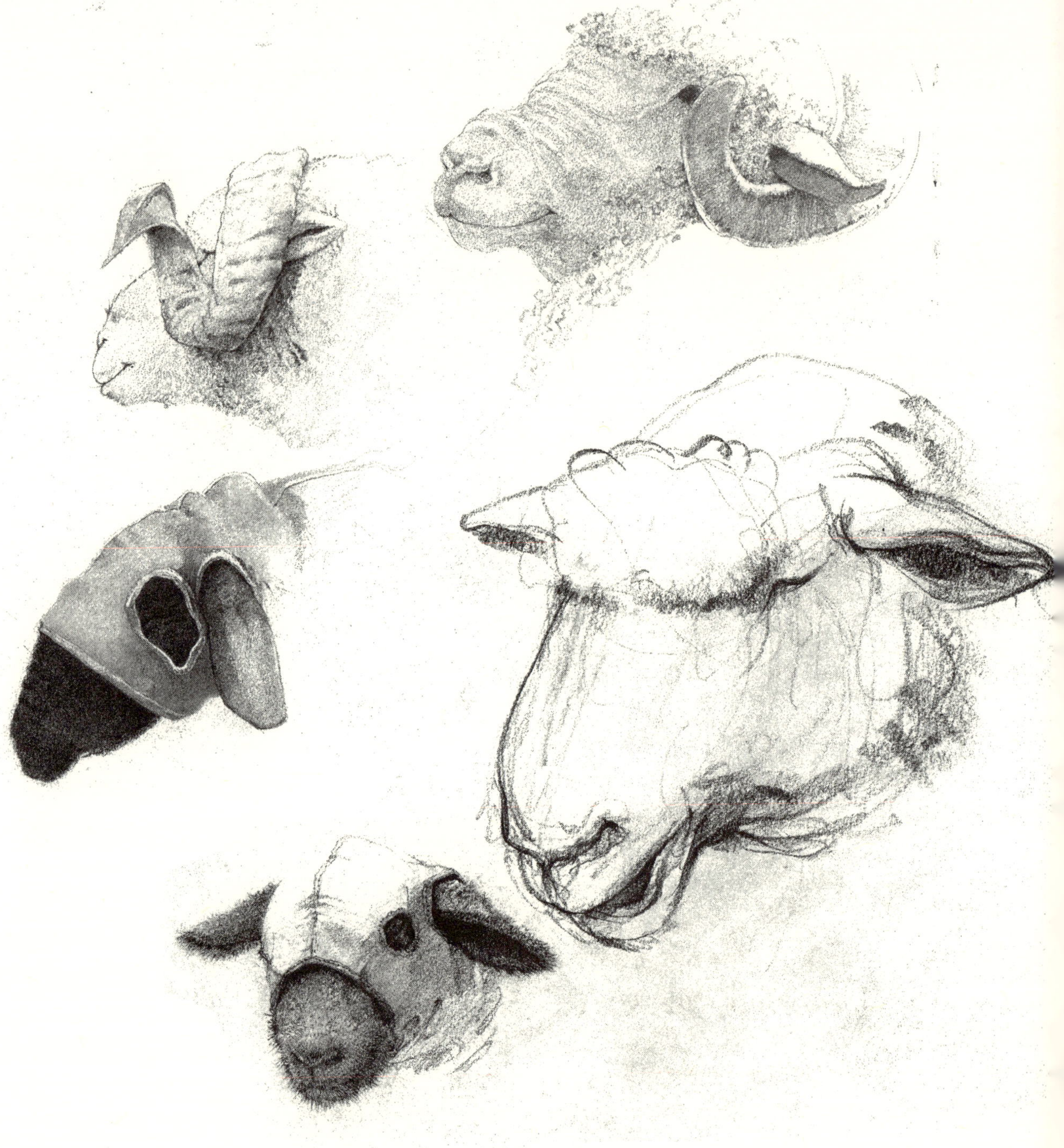

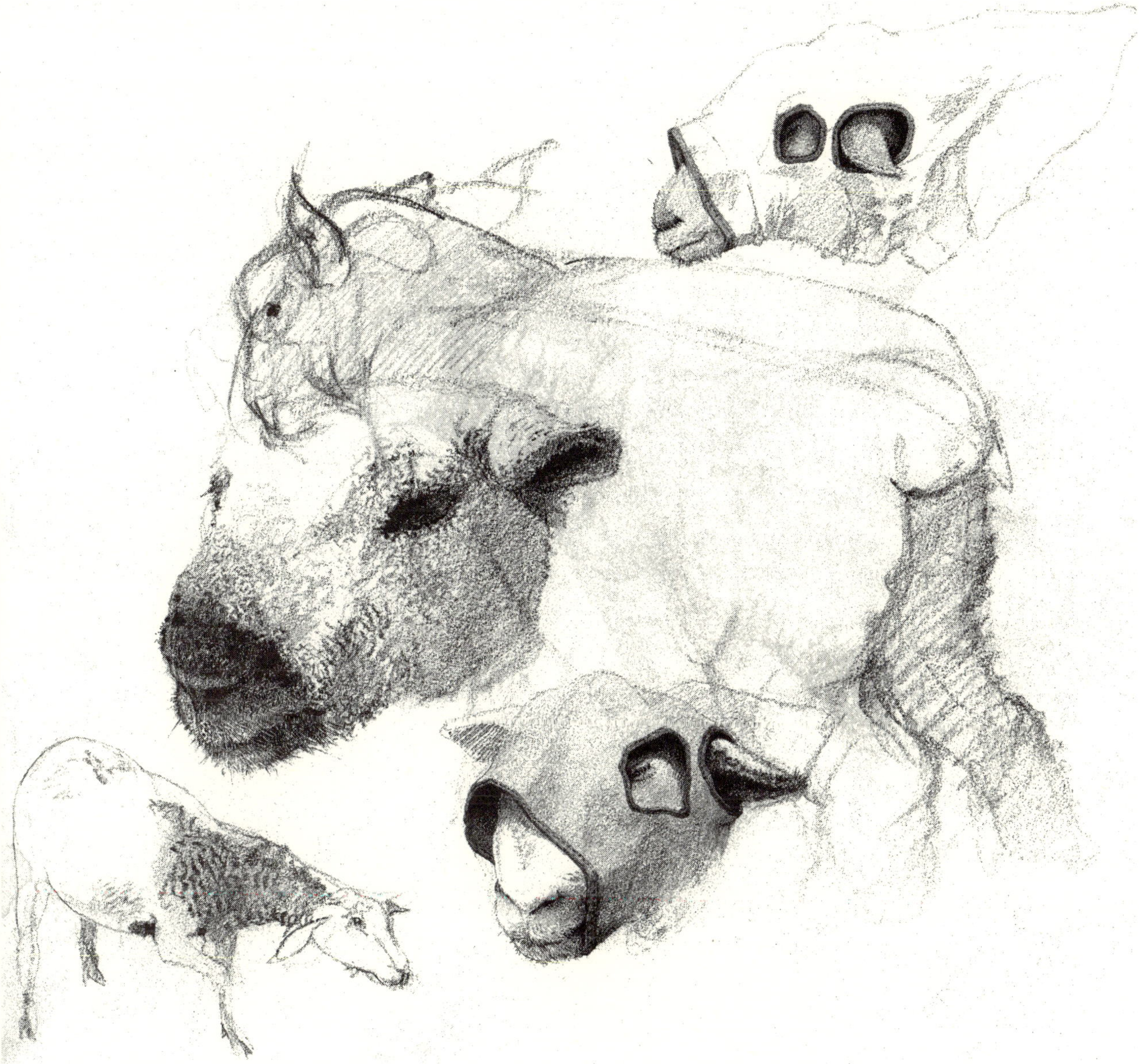

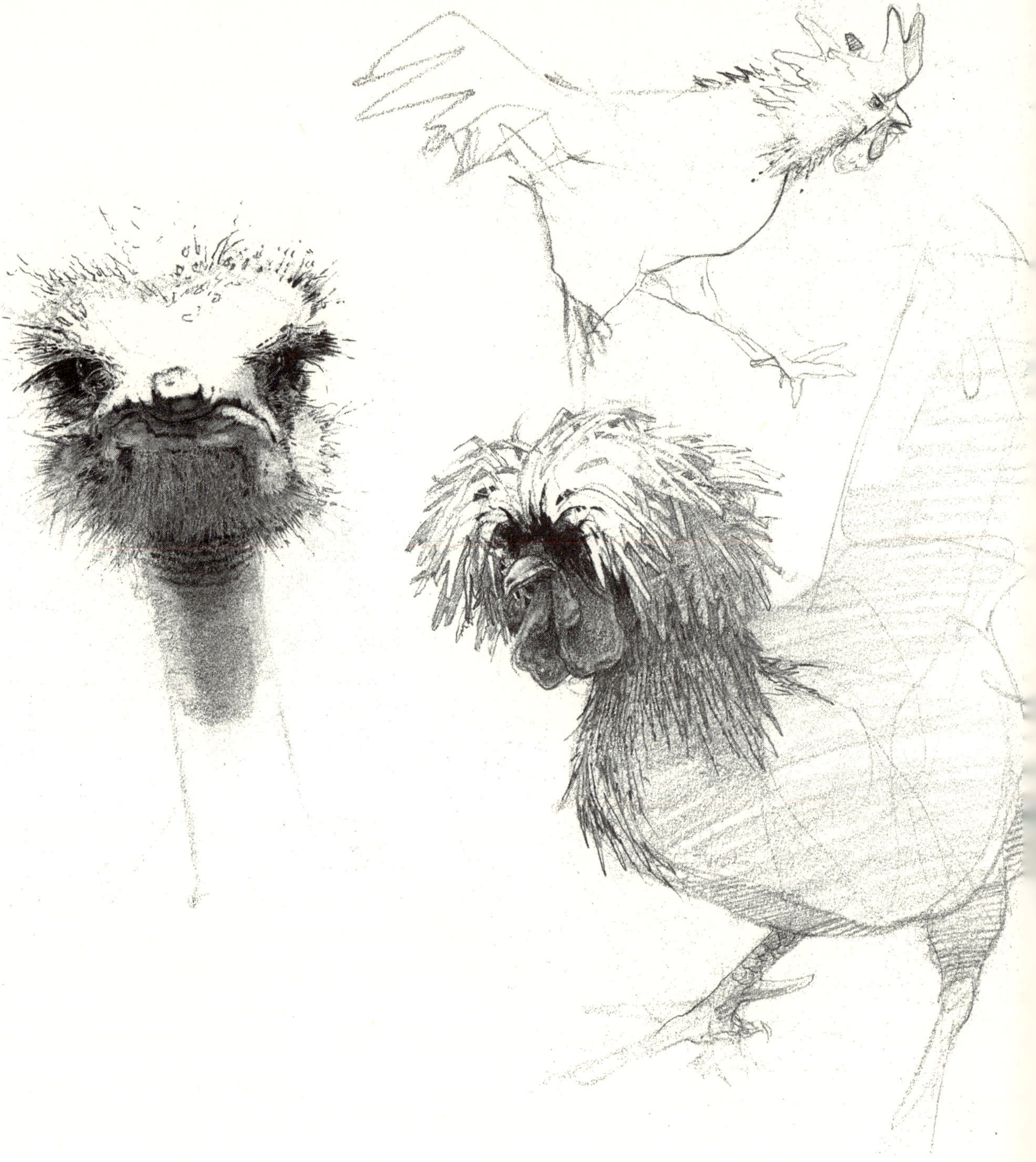

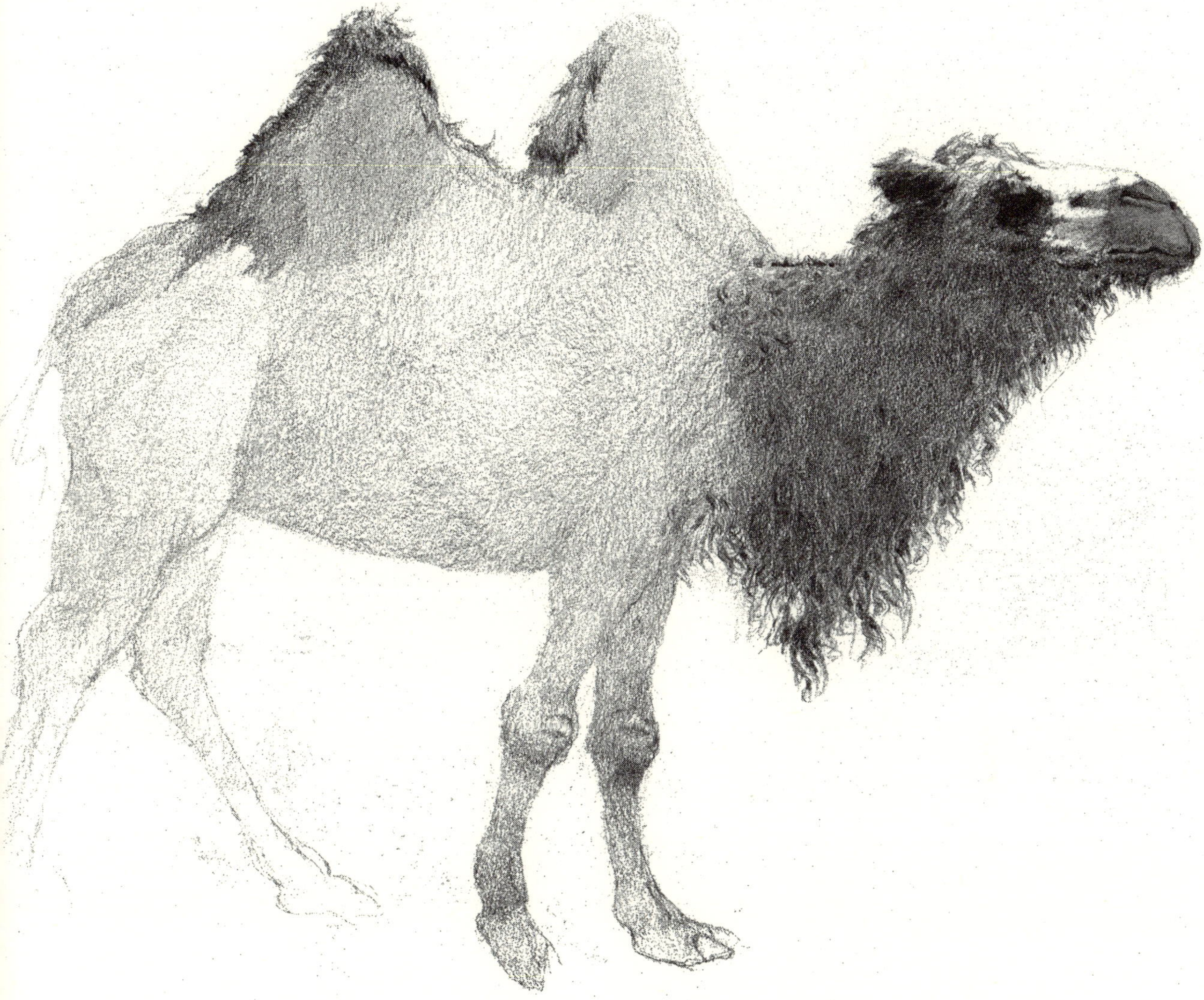

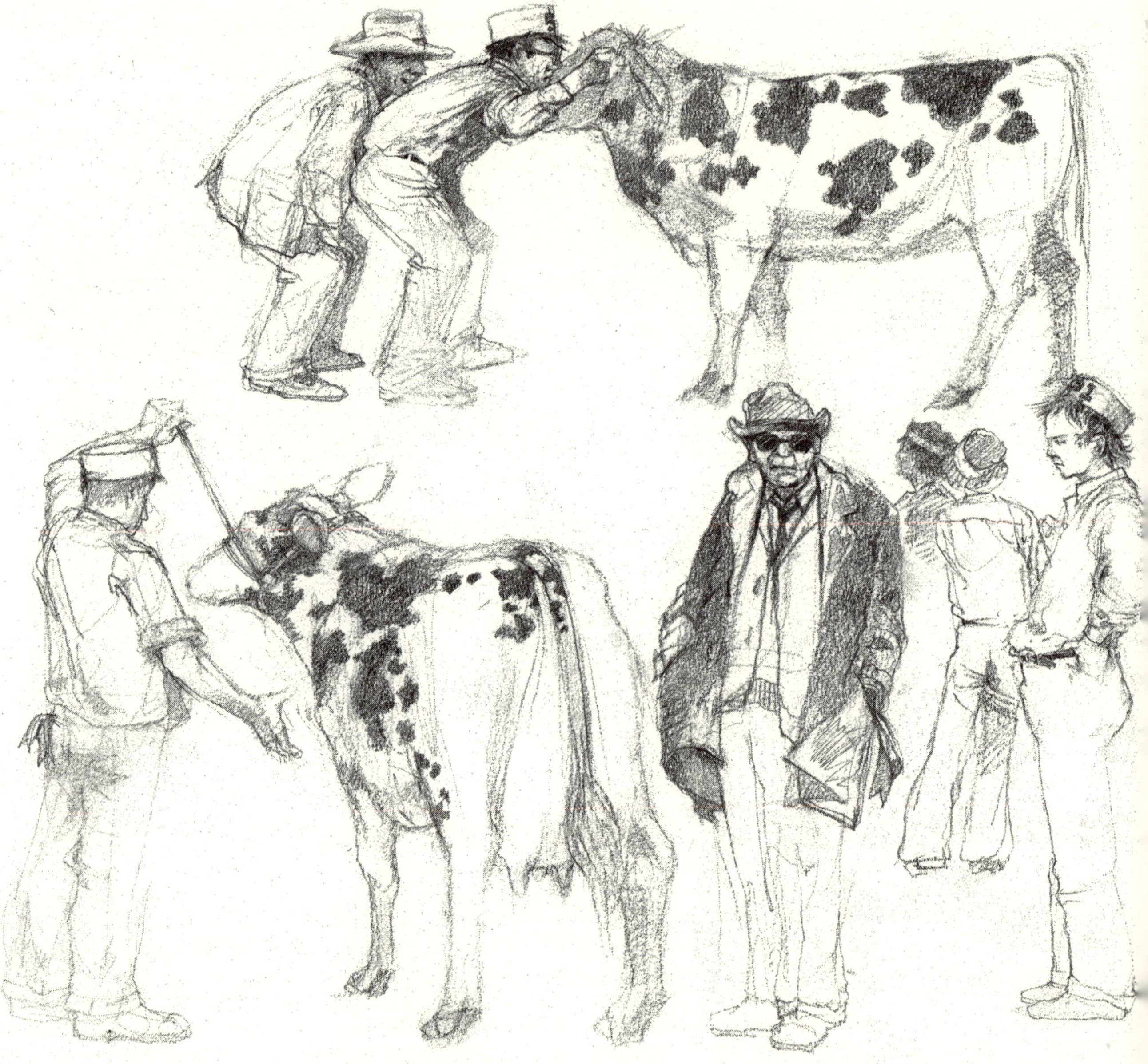

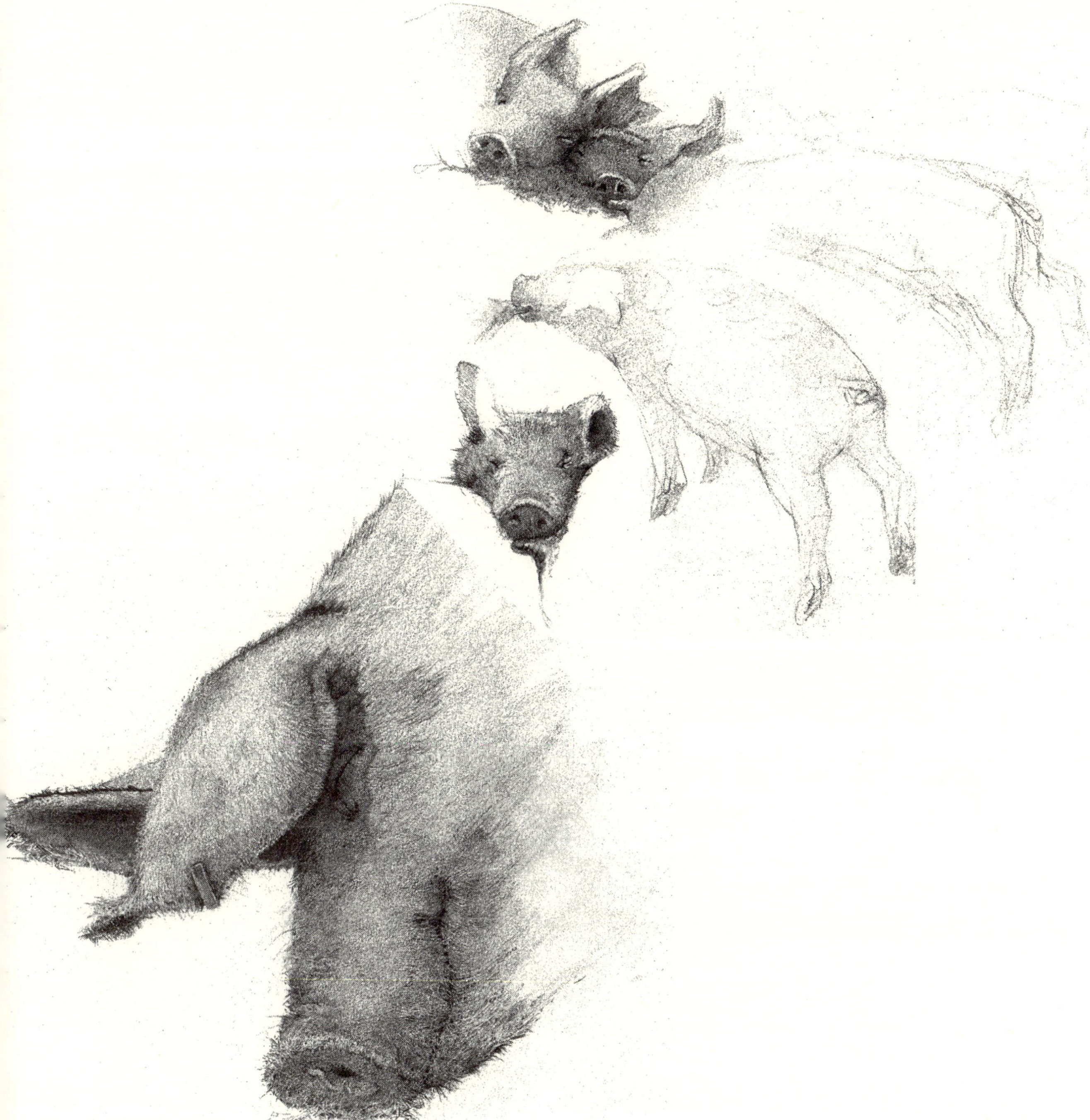

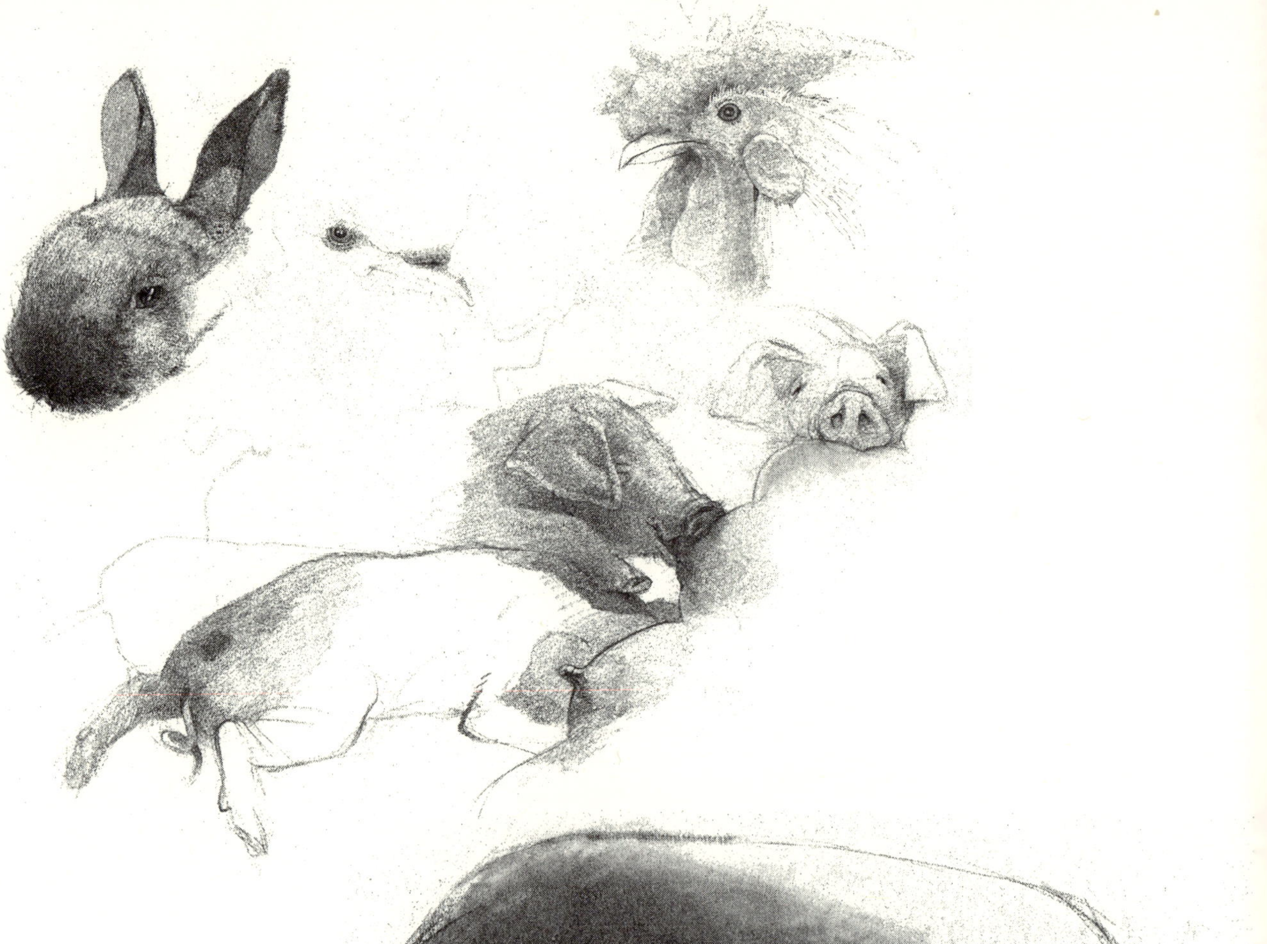

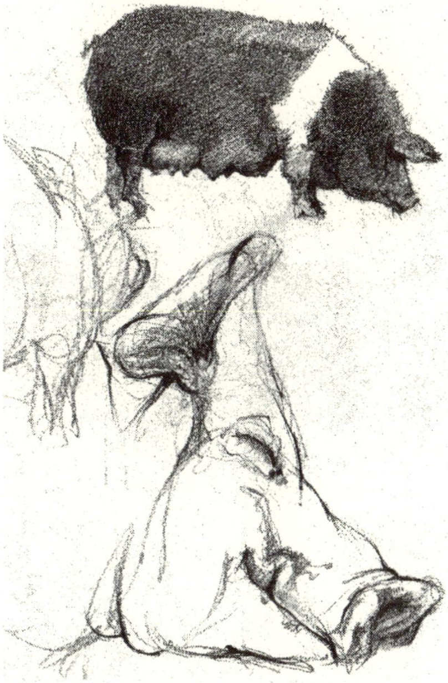

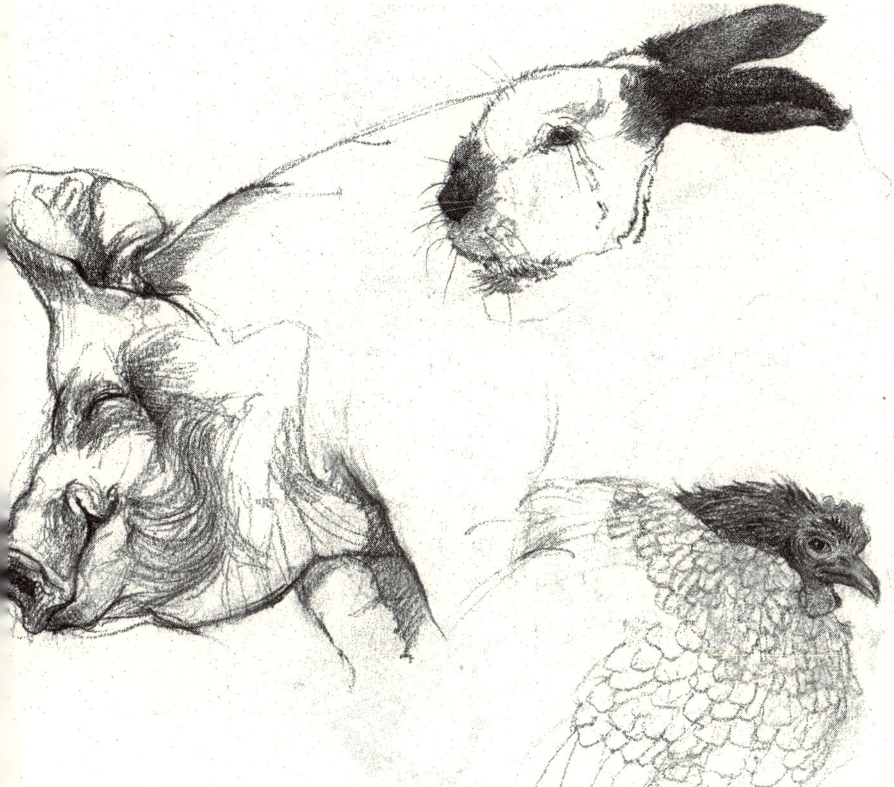

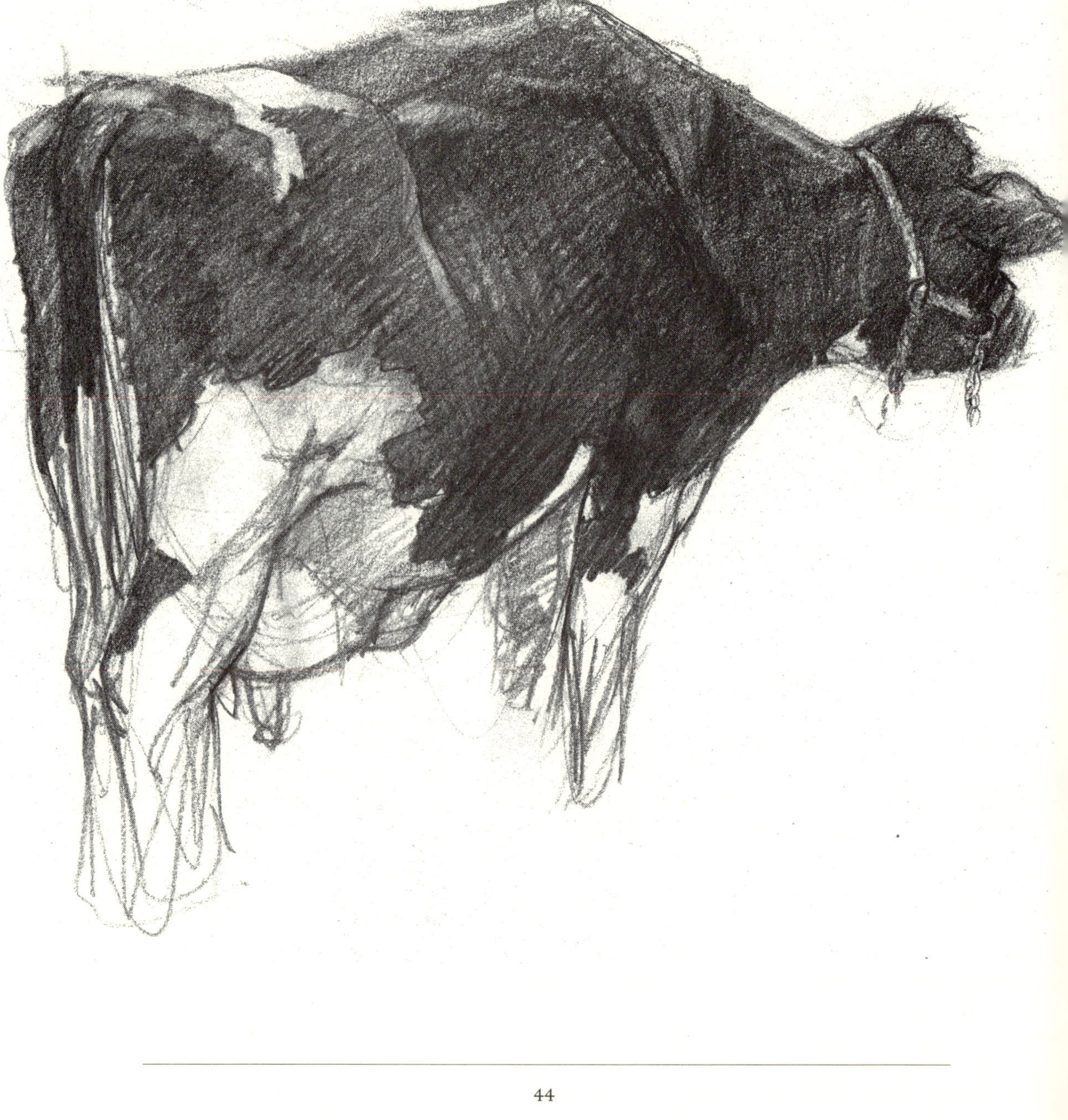

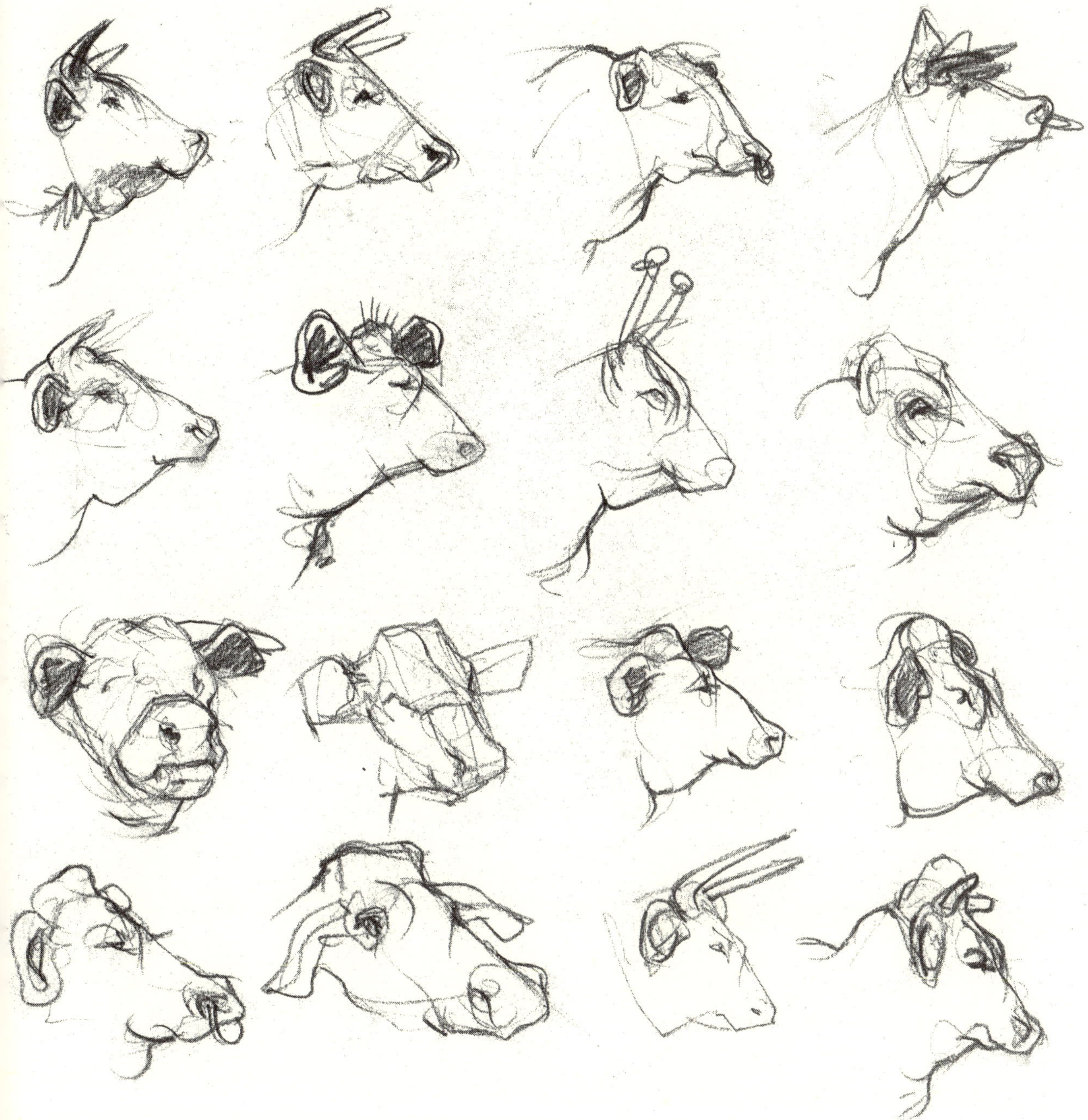

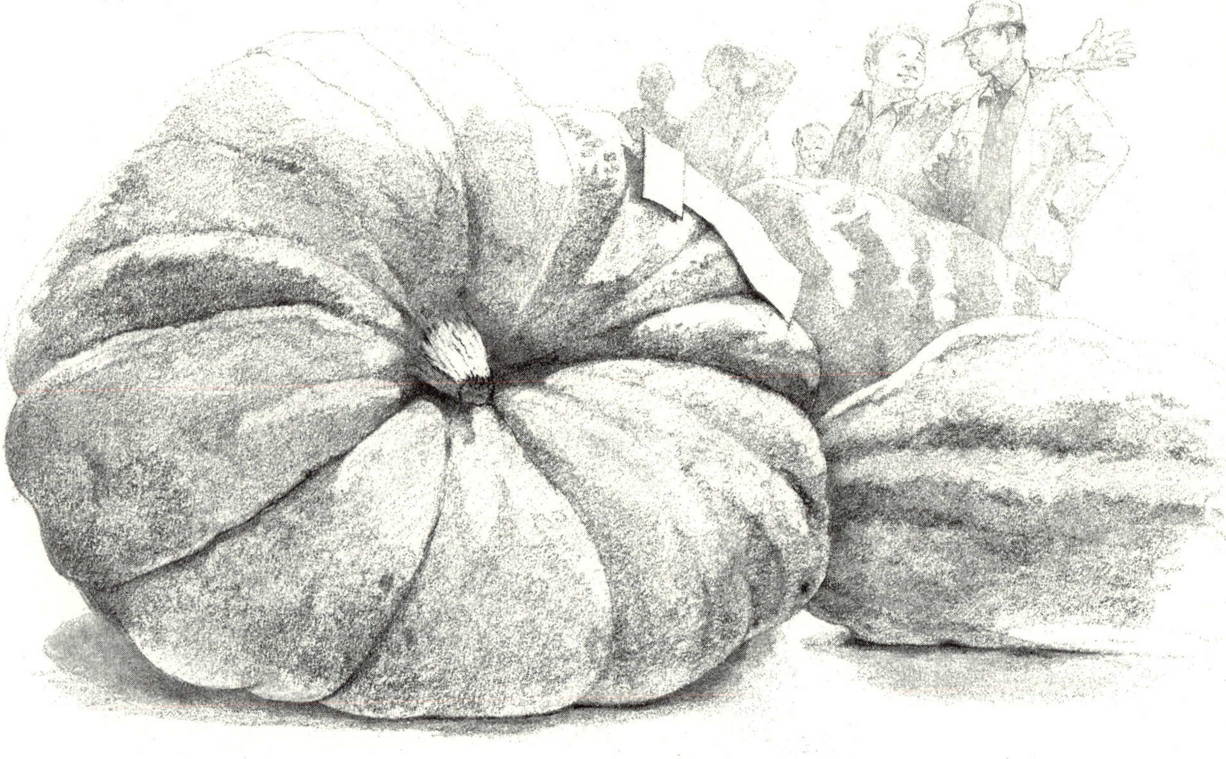

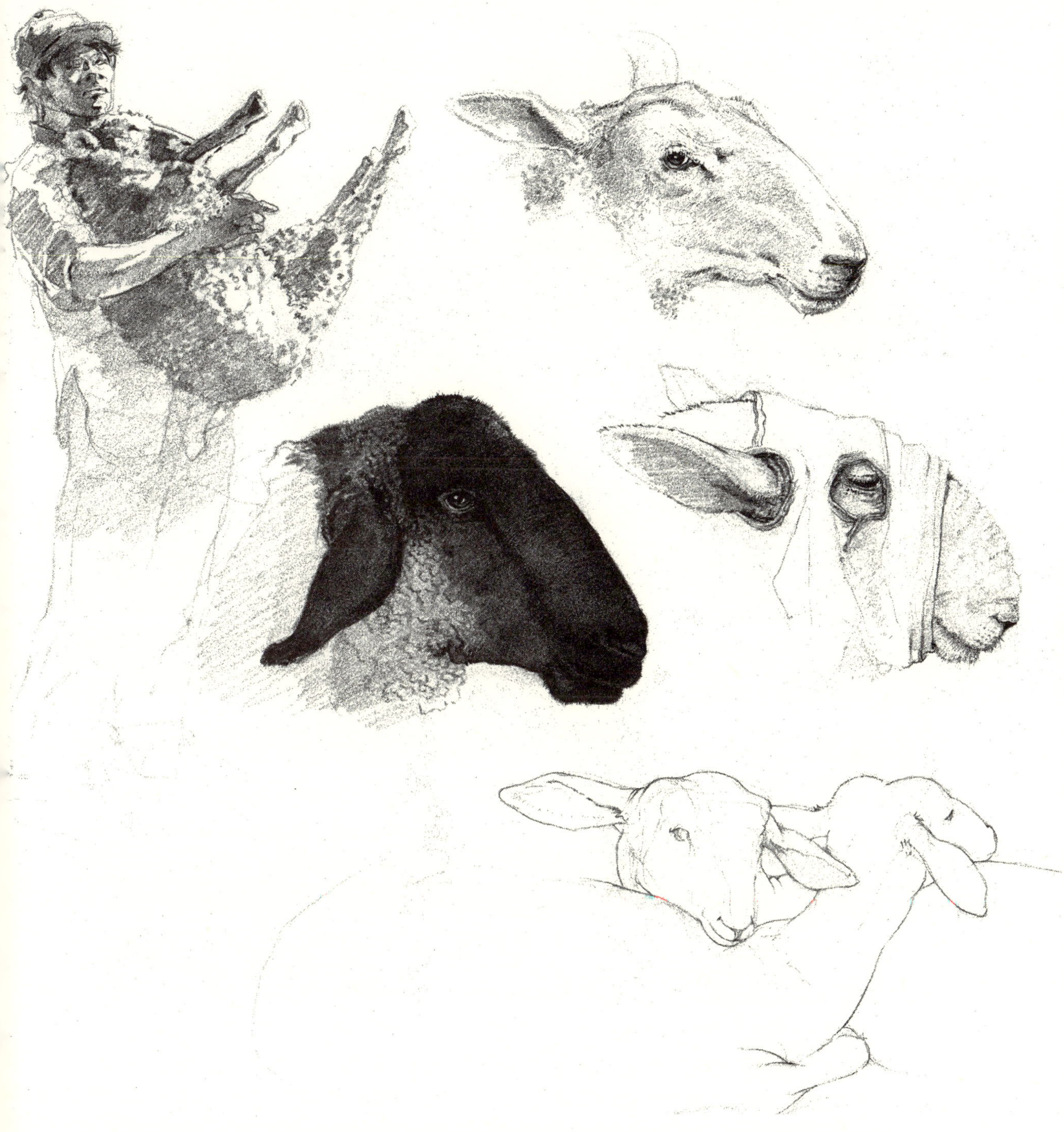

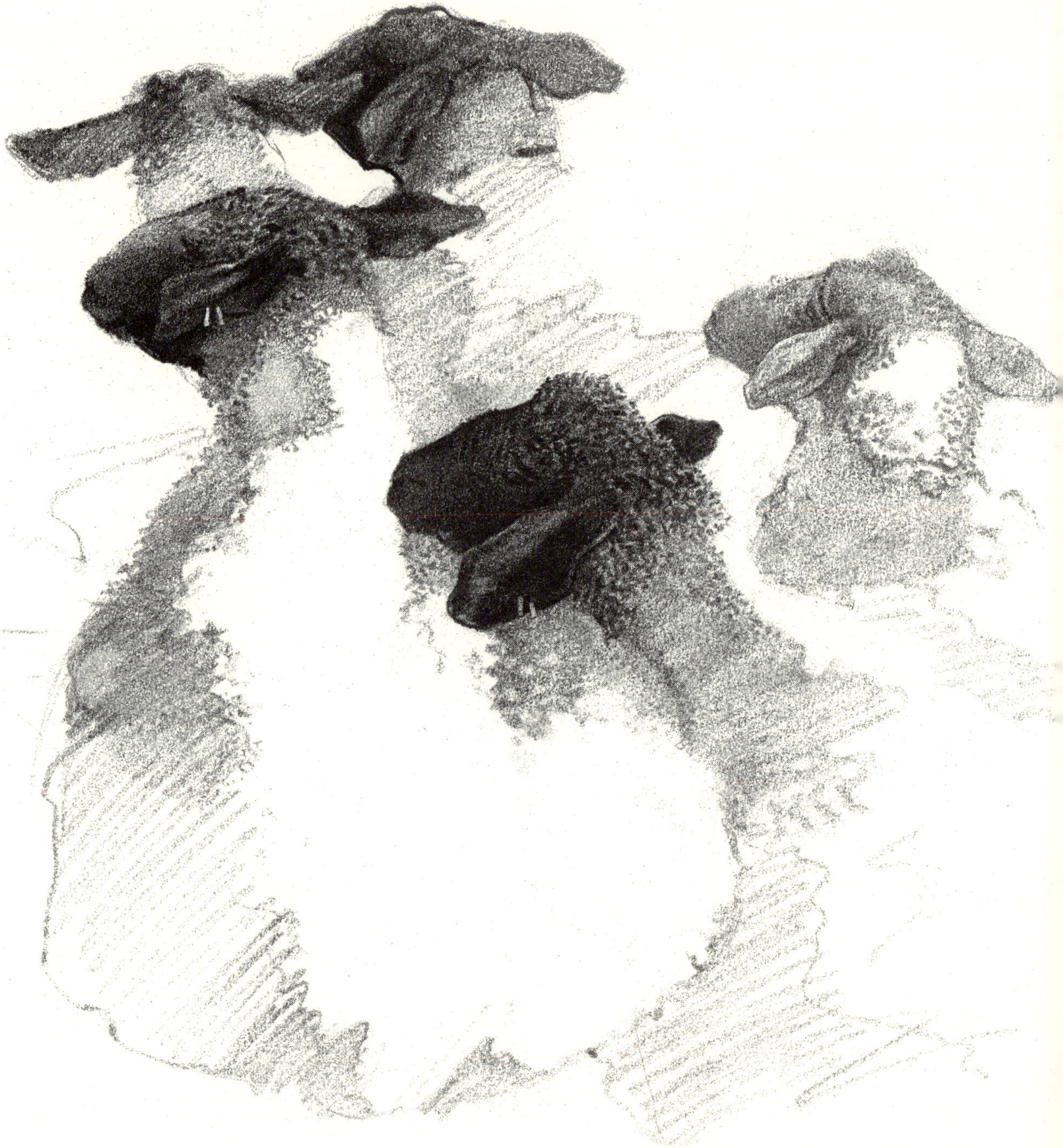

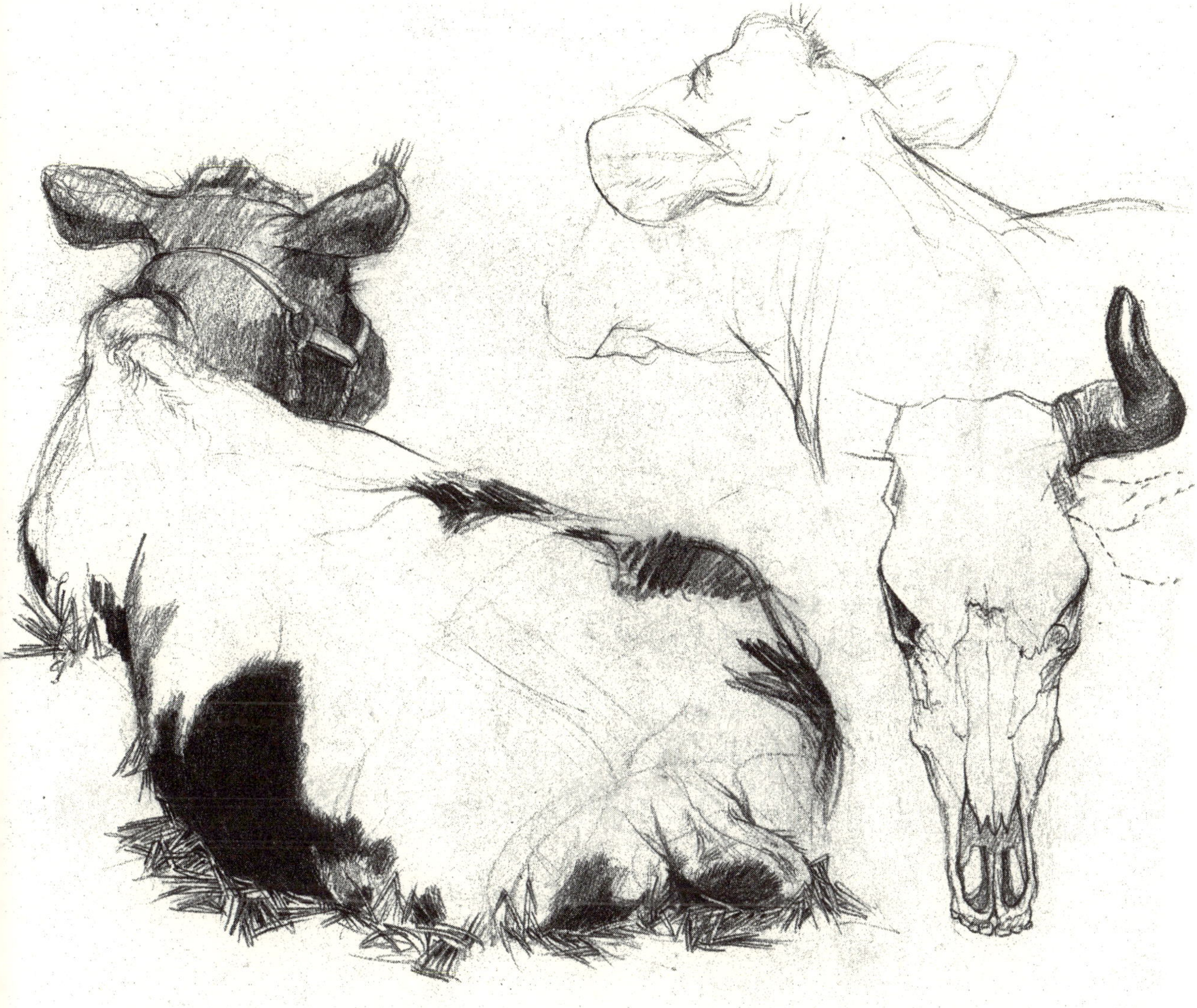

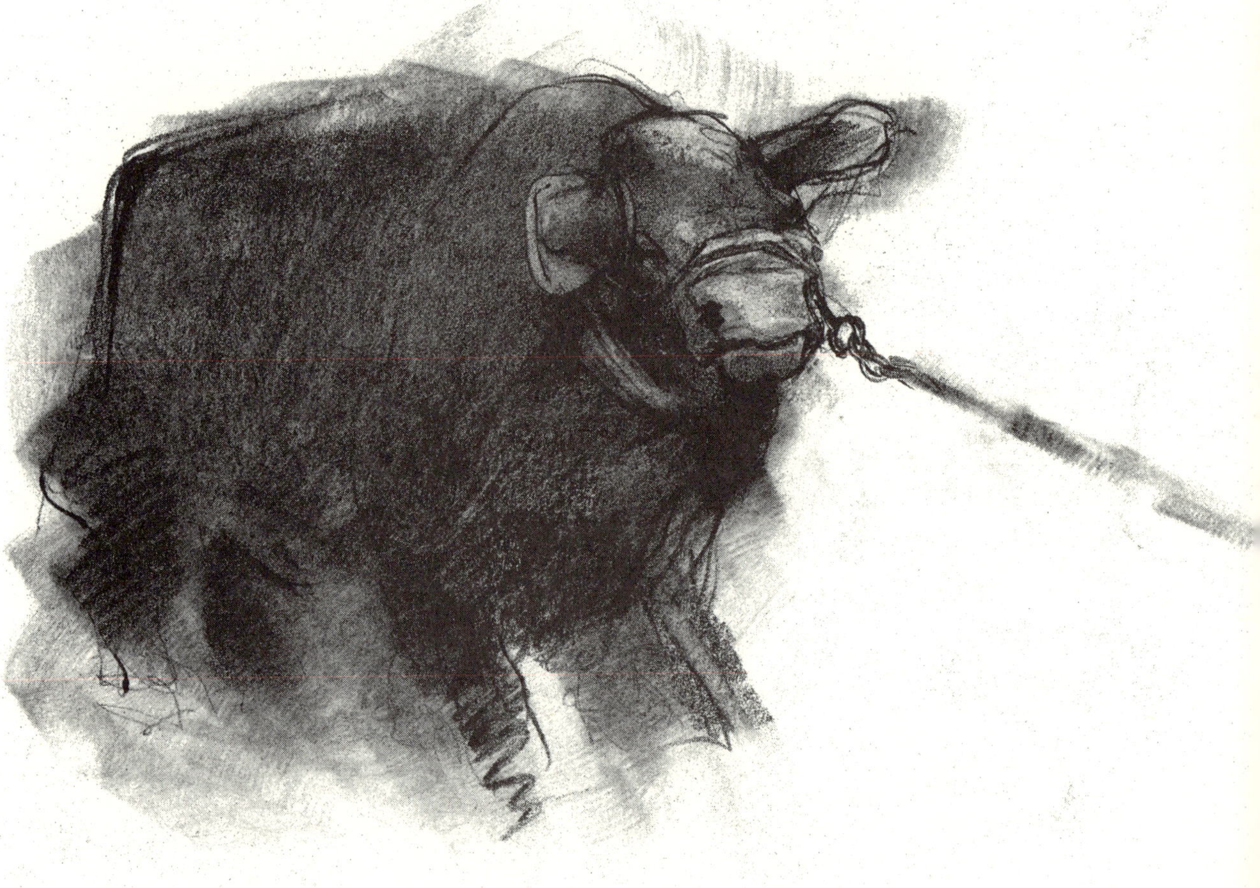

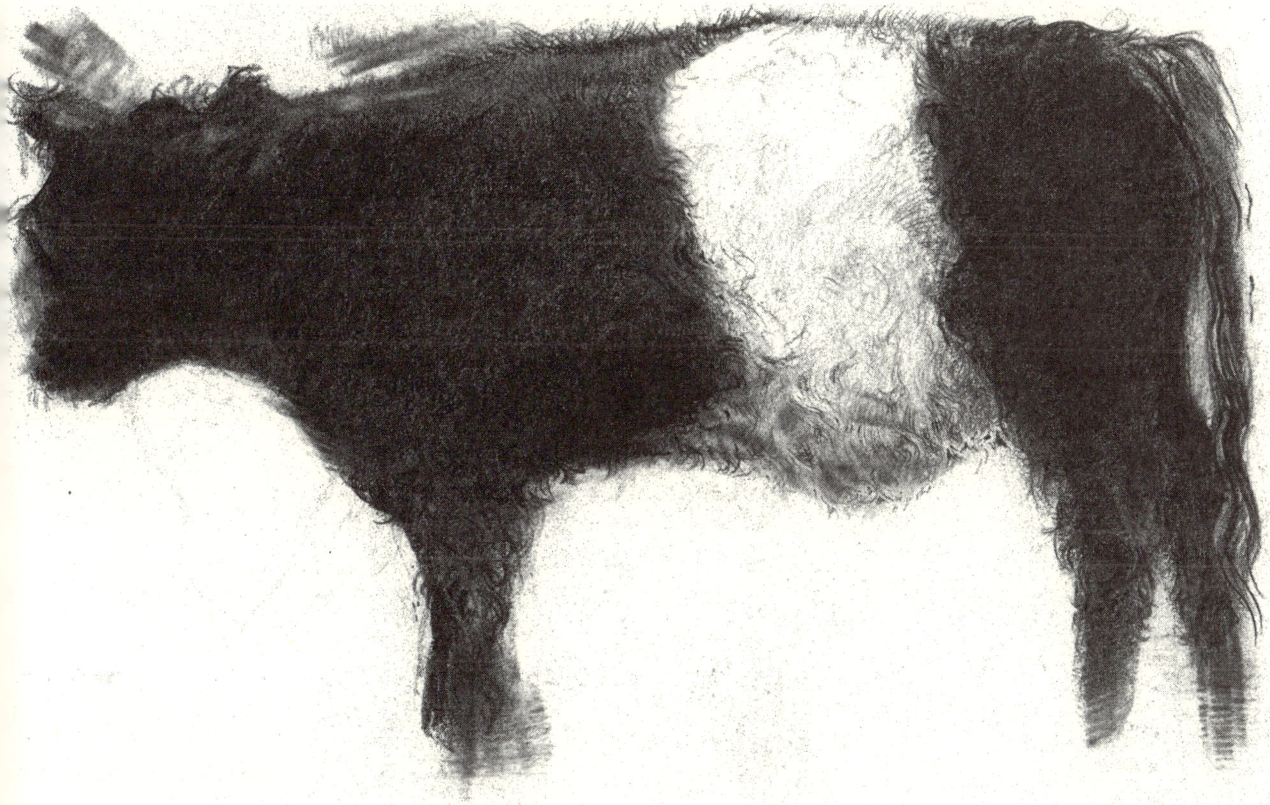

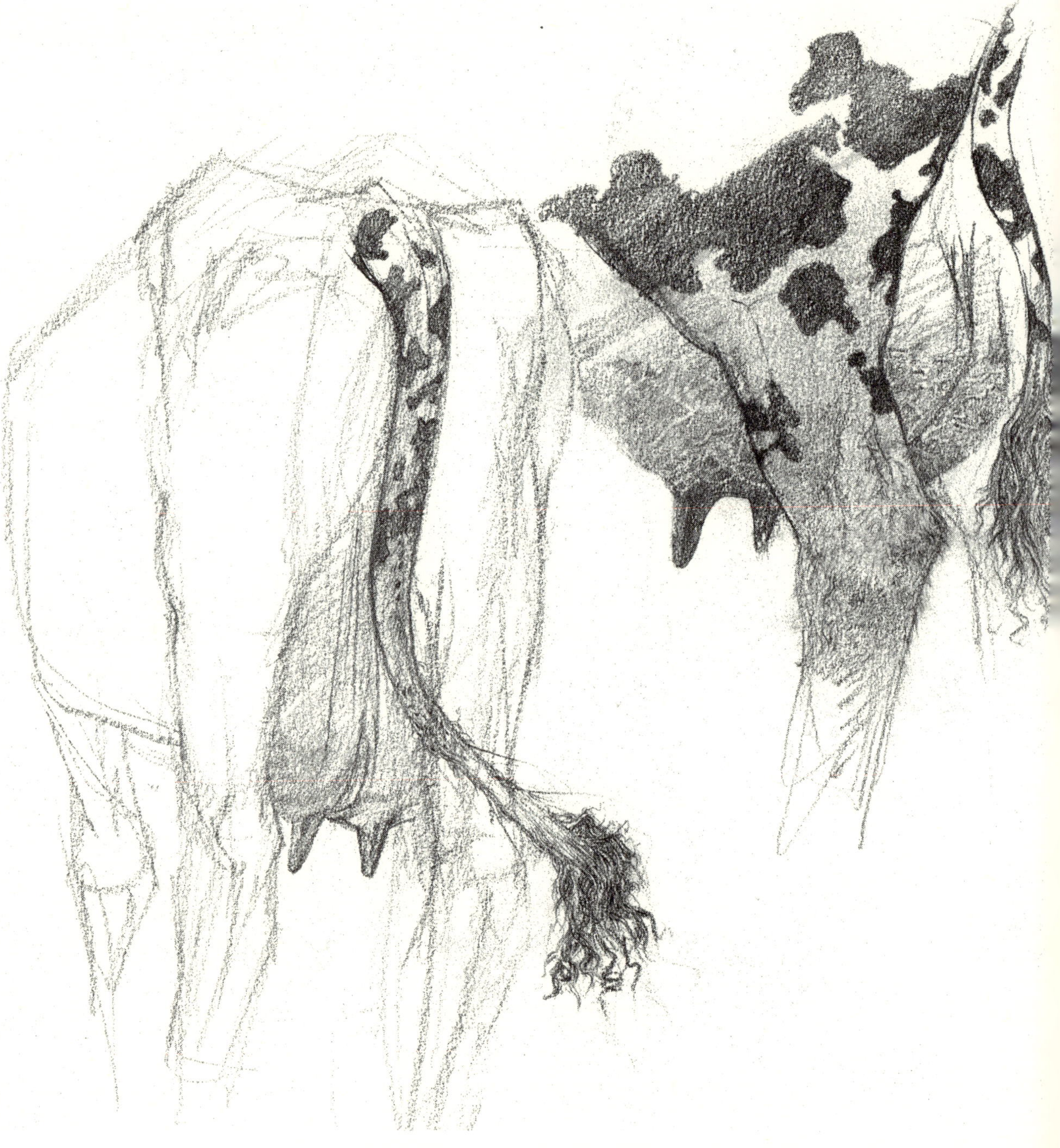

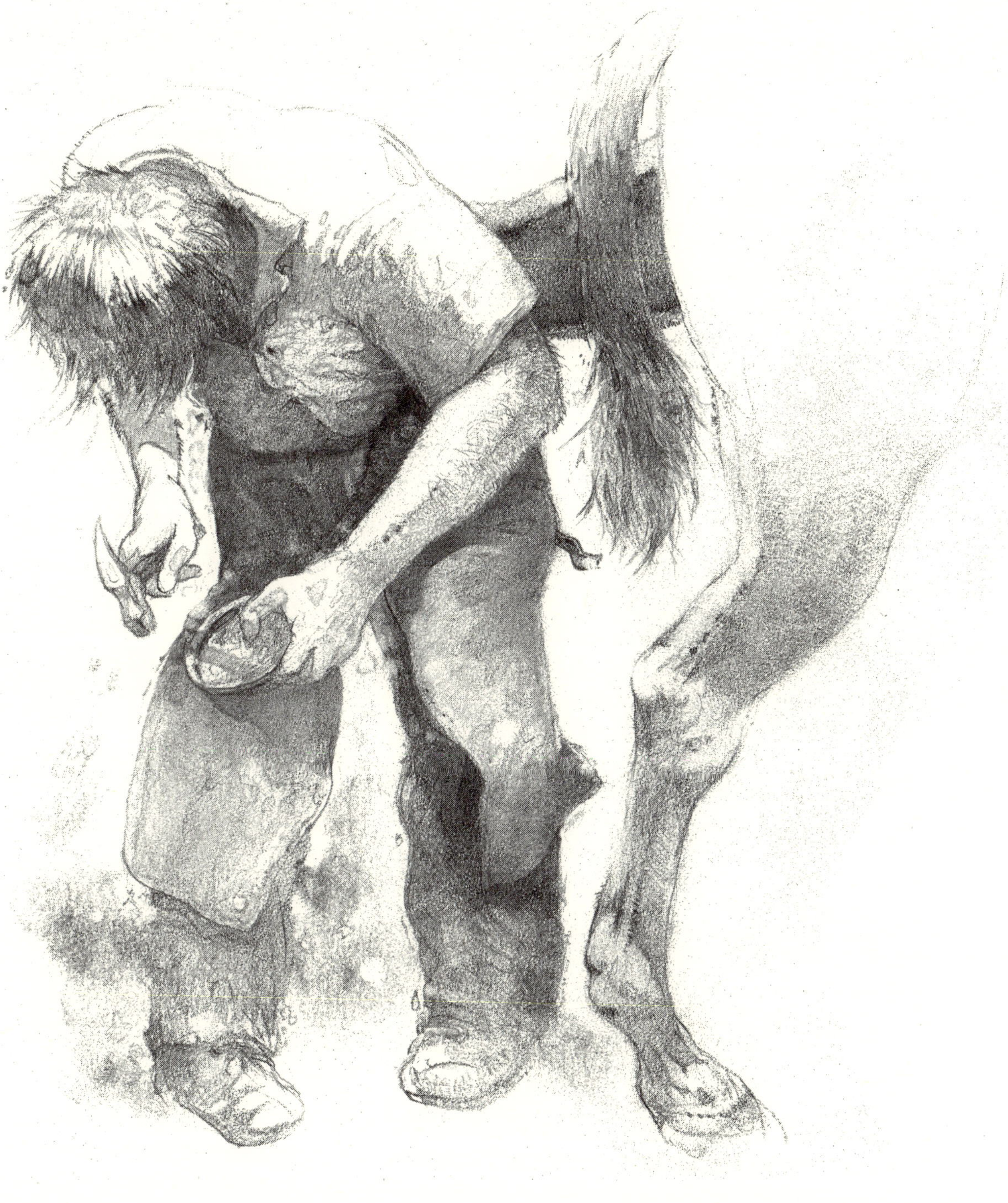

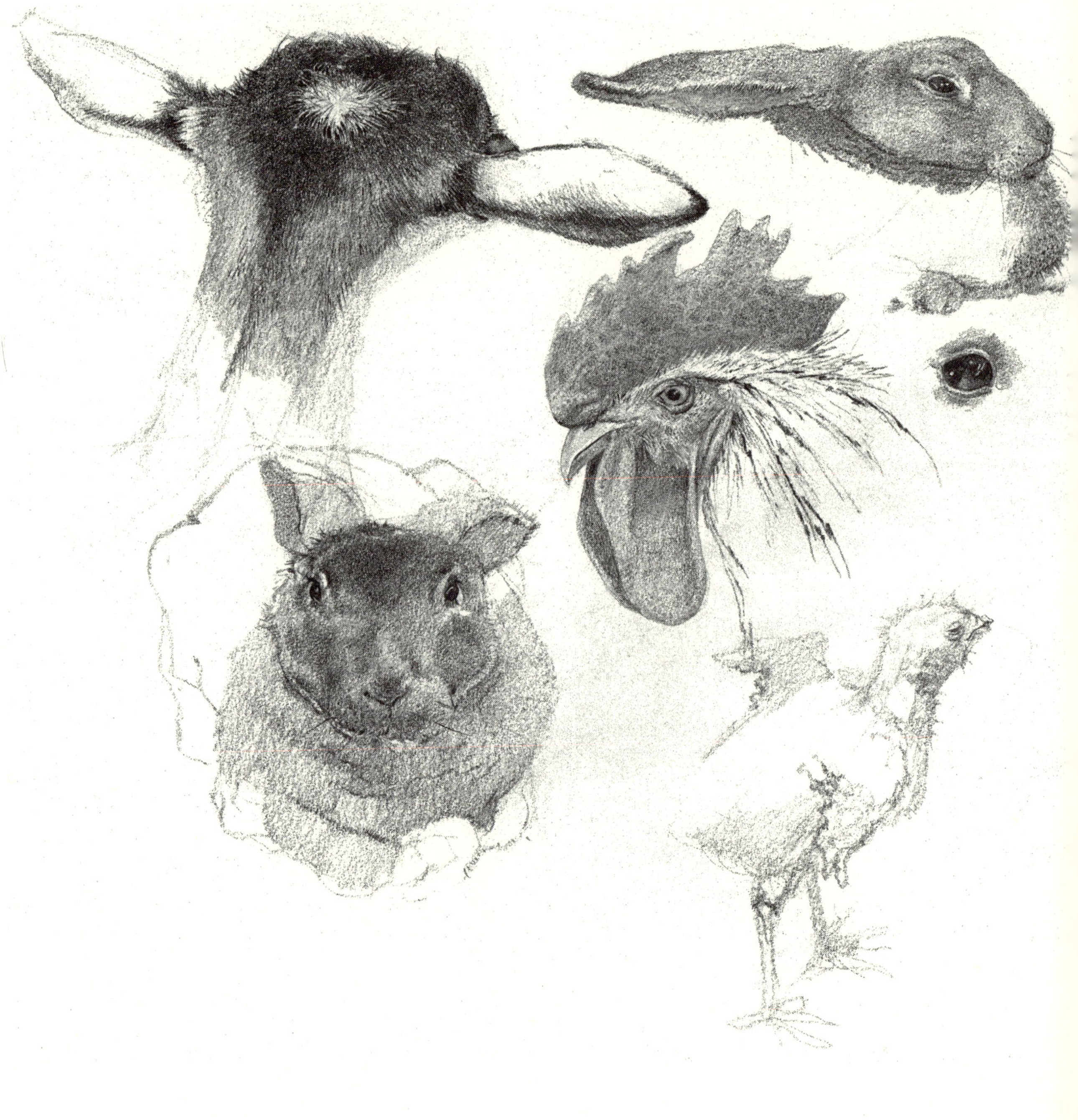

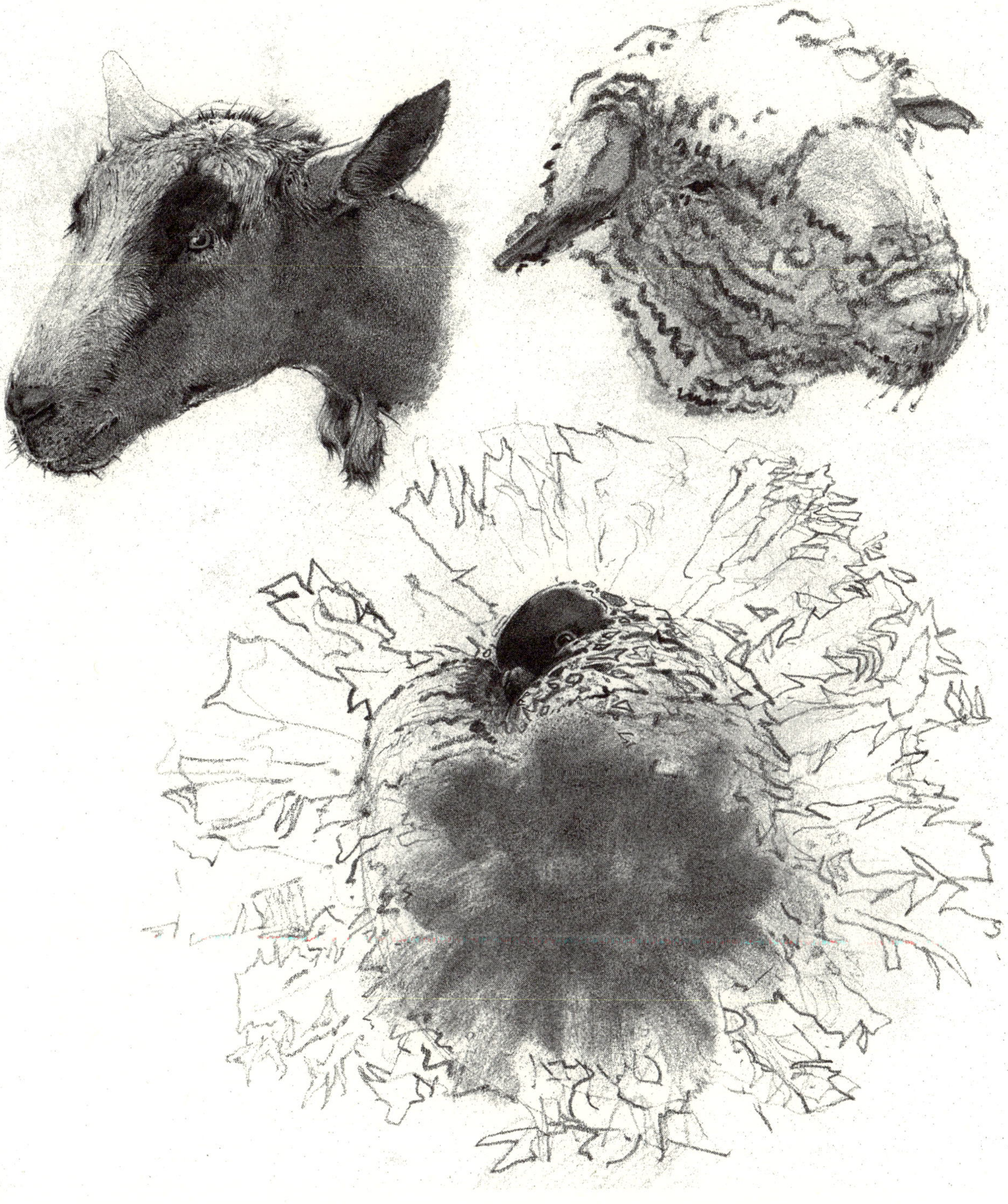

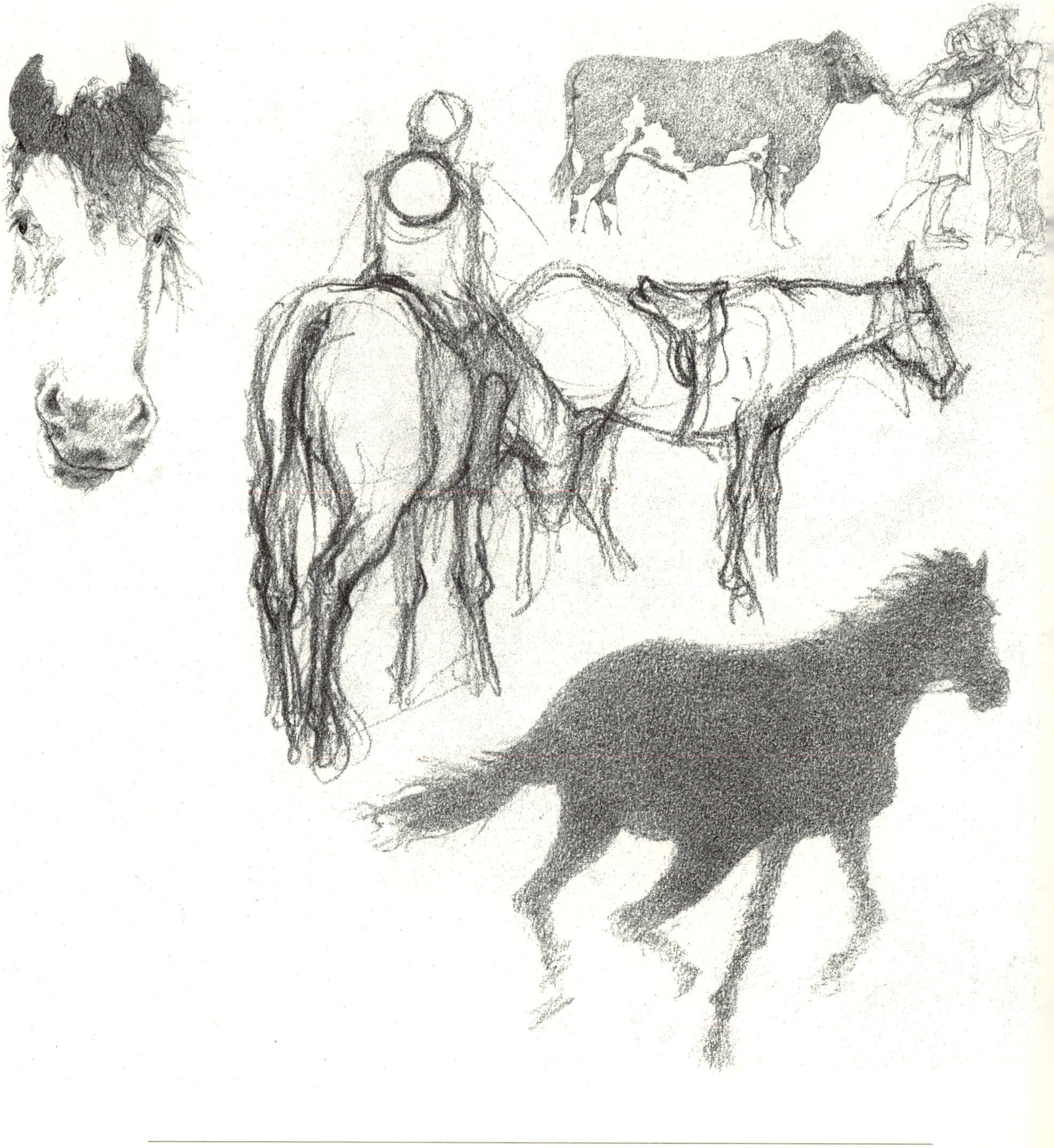

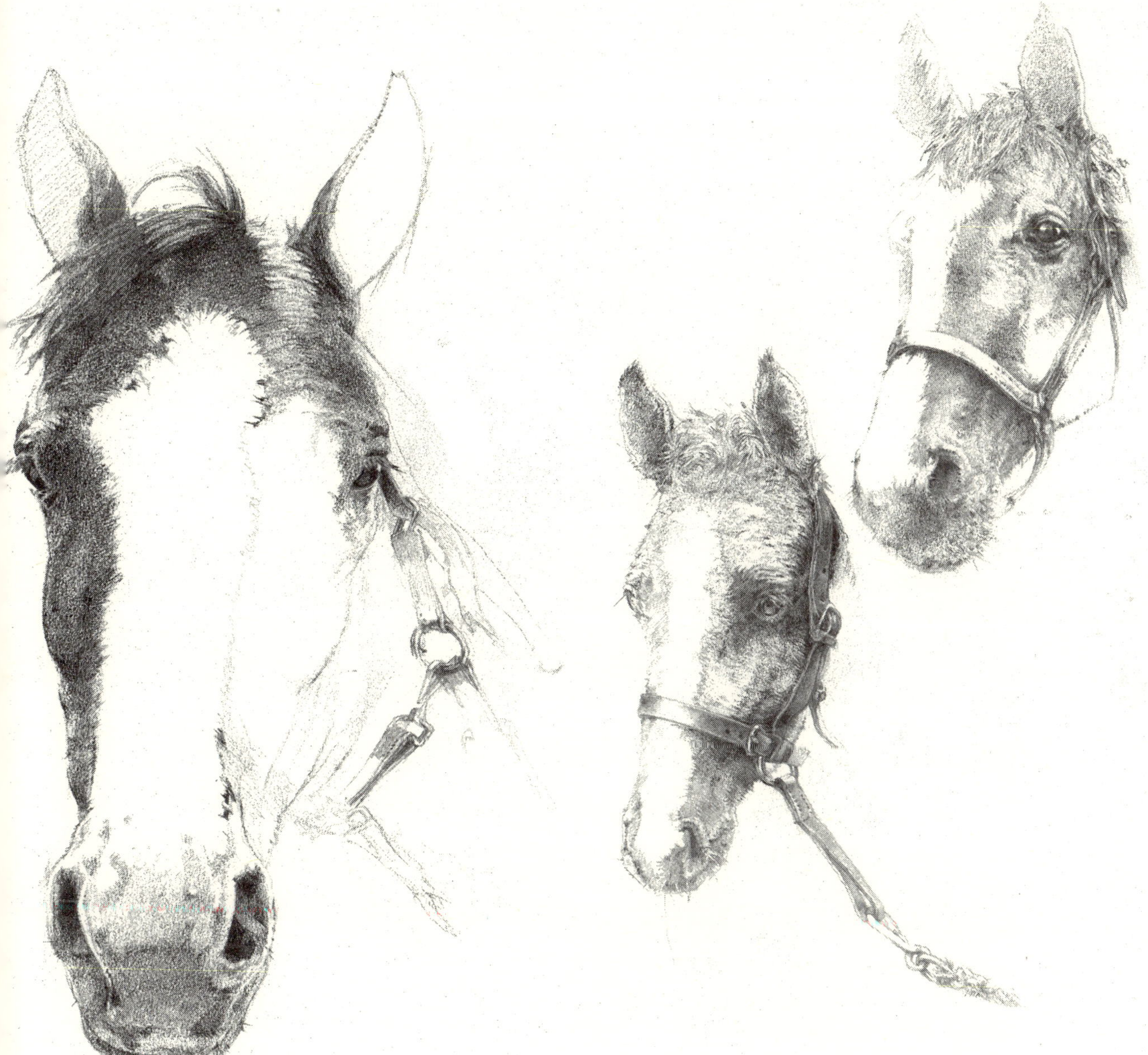

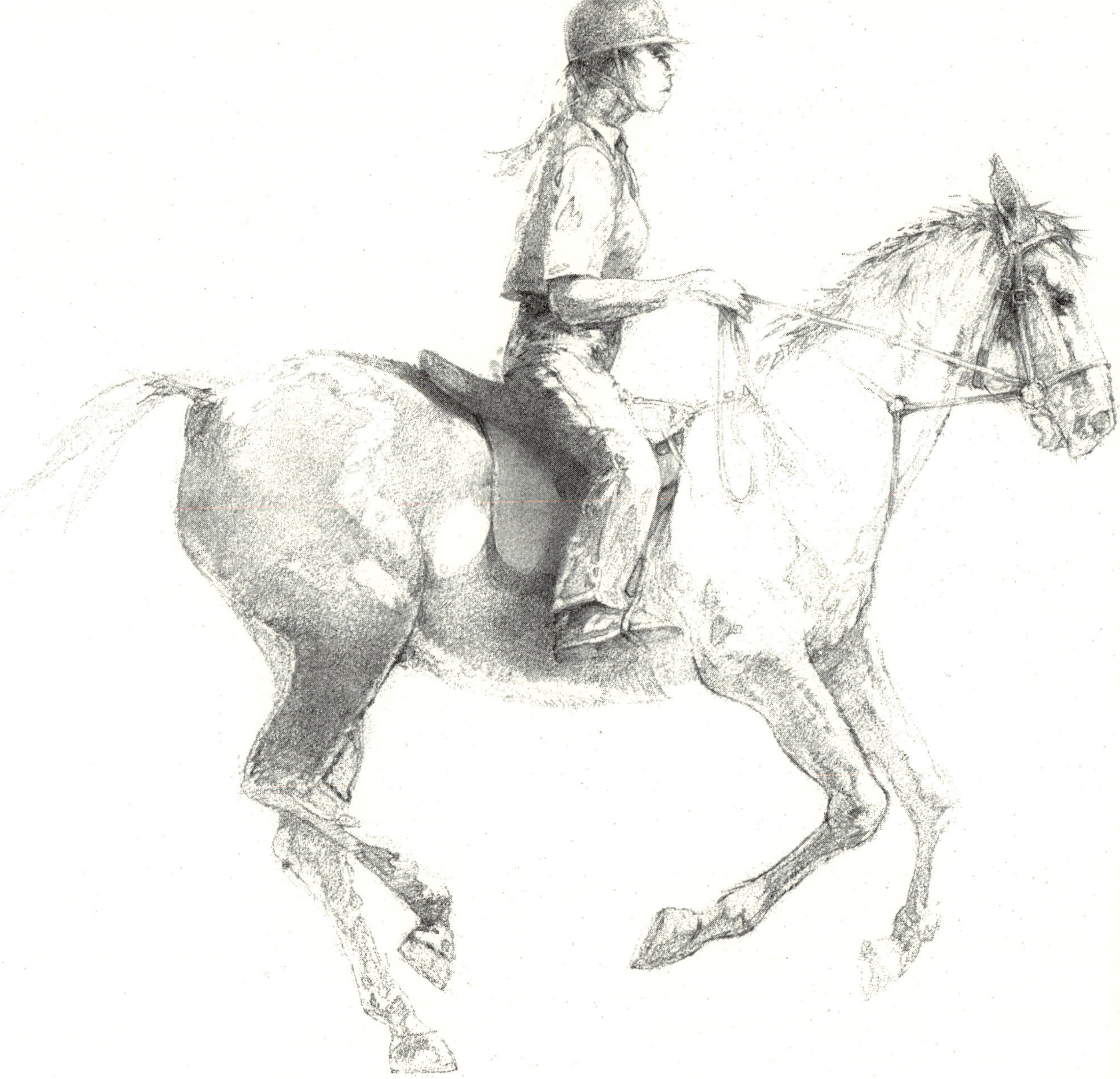

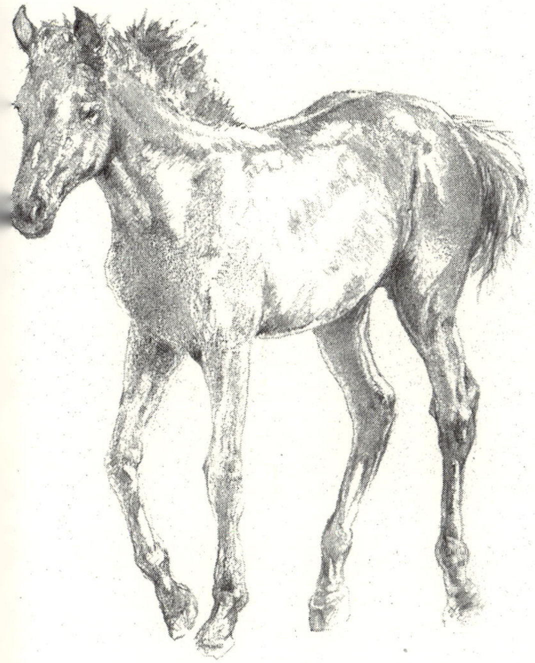

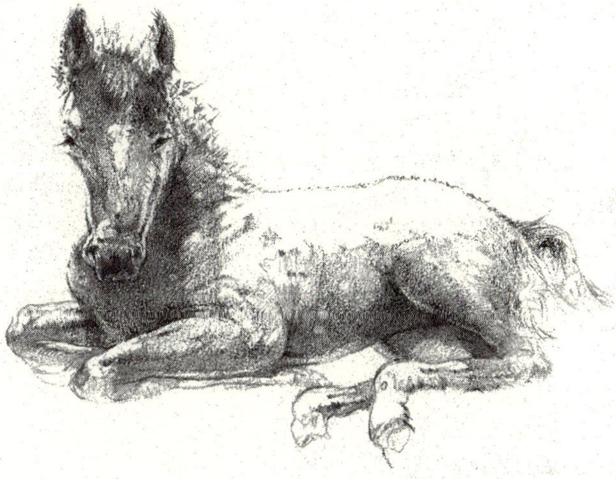

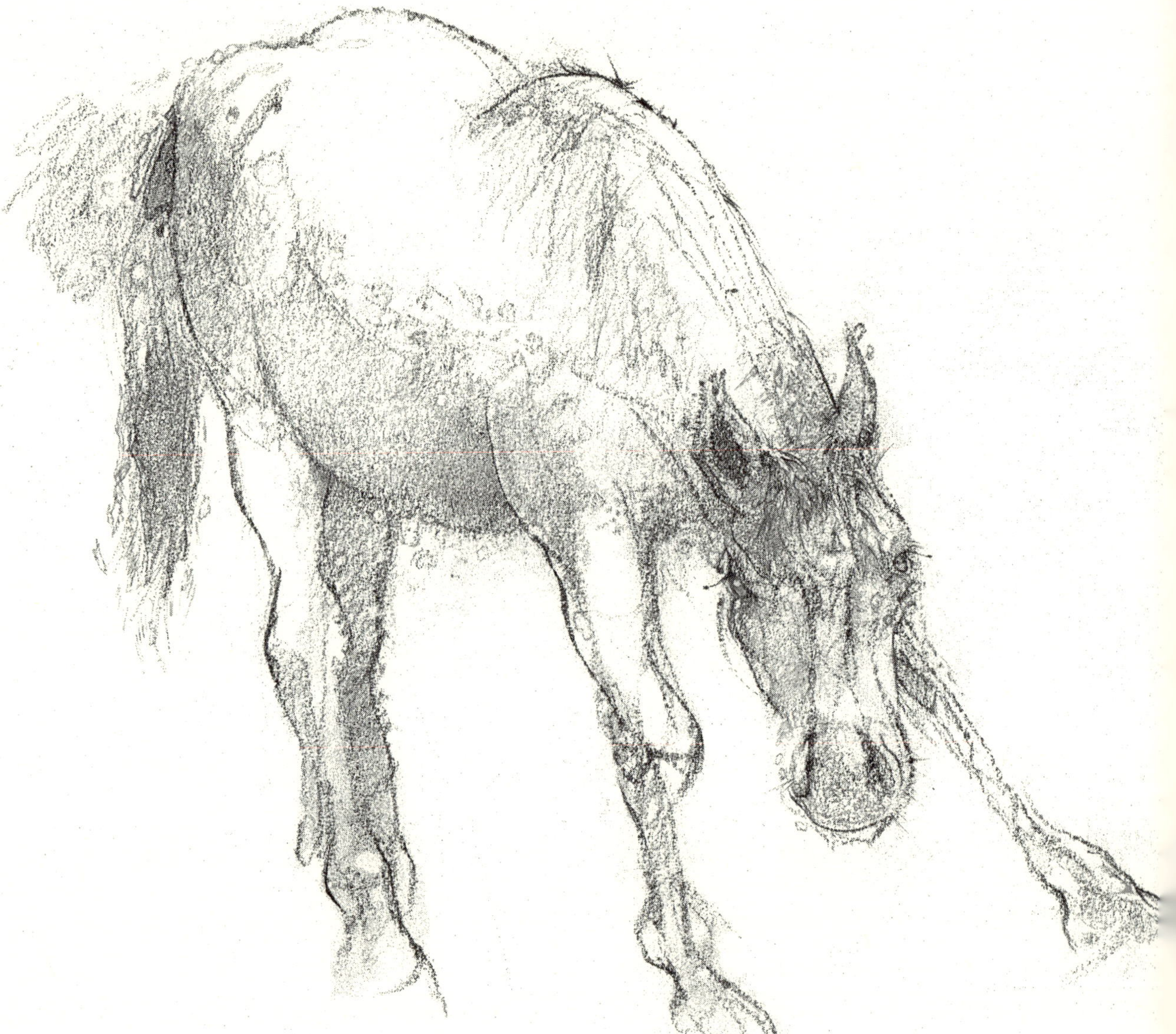

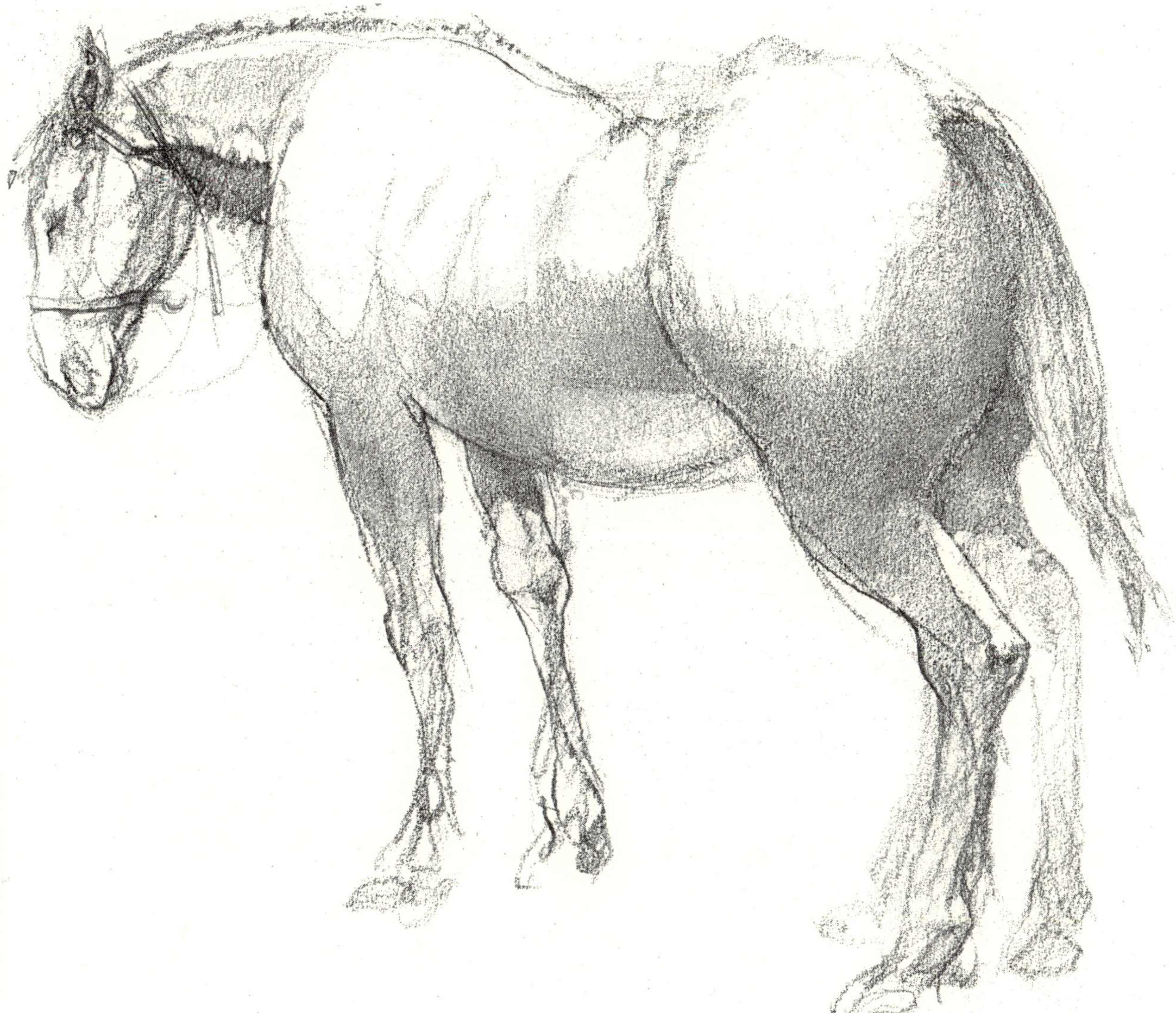

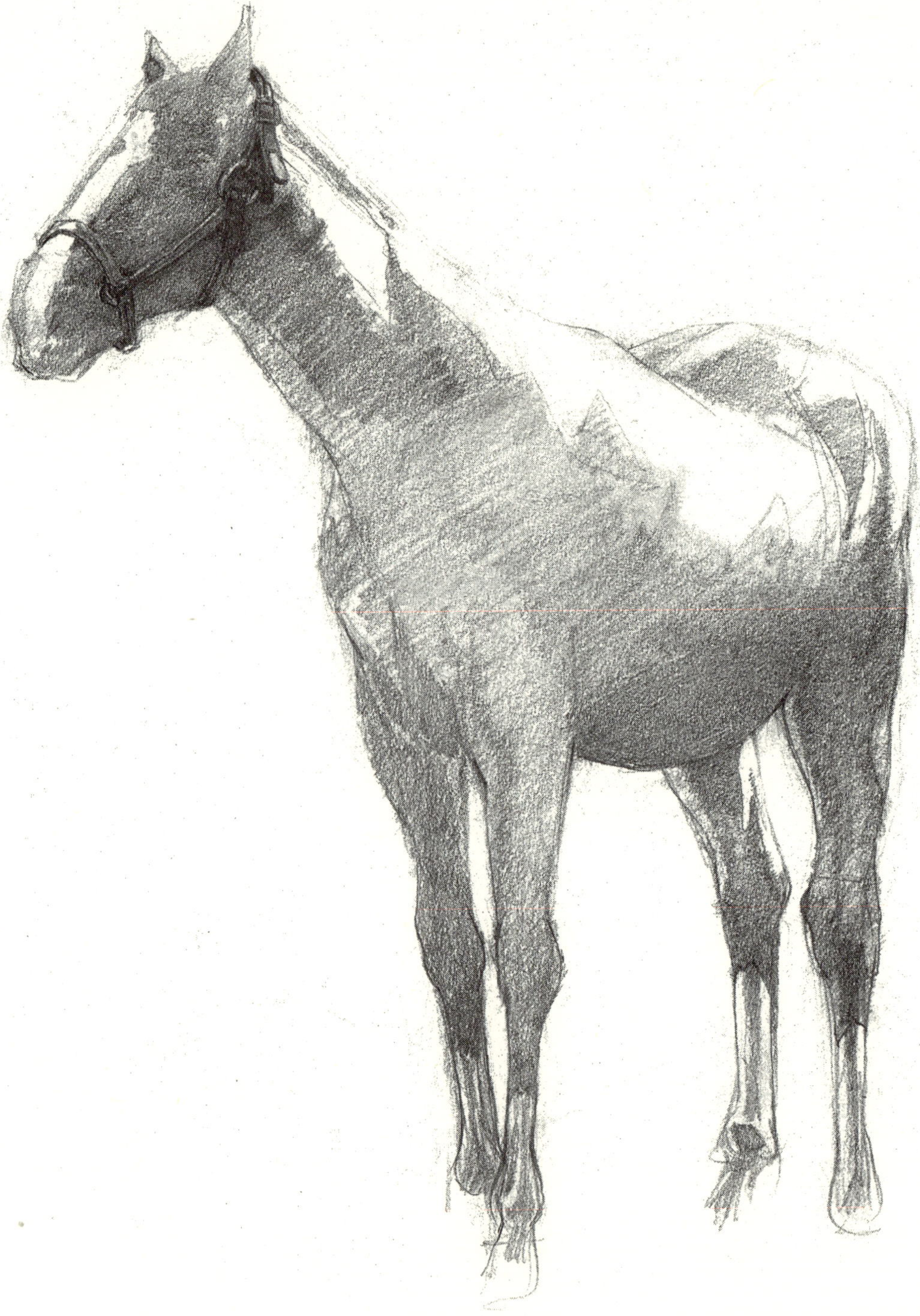

Introduction

TO ACTORS AND ATHLETES, THE OSCARS AND THE Olympics are the world's premiere events. To equestrians and farmers, it's the Royal — the largest annual indoor agricultural fair on earth.

Sprawling across 27 acres, the Royal is the only international exhibition to combine a livestock show, horse show, agricultural show and commercial trade show under one massive roof. During 12 days each November, 350,000 visitors and competitors, plus 10,000 livestock, cram the cavernous Royal Winter Fair site on Toronto's waterfront. And the mixture is amazing.

Equestrian teams arrive from around the globe (from Chile and Germany to Spain and Peru) to compete in the horse show — which, with 70,000 spectators, is the largest in North America. Buyers come from 35 countries. More than 40,000 school kids clatter in by trolley. Up to 500 international journalists are accredited. And over 800 volunteers flood the fair to assist the permanent 15-person staff.

Meanwhile, exhibitors pour in from every province, bringing 3,600 cattle and 1,300 horses (which eat 20,000 bales of hay), plus 2,000 birds, 600 goats, 400 swine and 300 sheep — to say nothing of giant 680-pound pumpkins and 1,700-pound cheeses.

The exhibitors have included such celebrities as hockey's Bobby Hull (who proudly showed his Hereford bull-calf, Powerplay), Ontario-born cosmetic doyenne Elizabeth Arden (who brought her entire stable of prized horses) and Harry Oakes of Niagara Falls, who (before being knighted, then mysteriously murdered in Nassau) sponsored the Hunter Stake.

There was also, of course, Alberta rancher John Boake. While feeding his cattle en route to the fair, Boake became an instant legend by toppling off the train. (The engineer, who backed up 20 miles to retrieve him, remains unsung.)

Other famed names appear only in the ring. Arthur Godfrey used to perform the lavade with his Palomino stallion, Goldie. The RCMP Musical Ride (which originated in 1878 to fill the idle time of North West Mounted Police officers) often packs the Coliseum. Dwight D. Eisenhower once arrived in a George IV Phaeton drawn by four hackneys. And over the years, the British Royals have often popped by in phaetons of their own.

Since it first kicked off in 1922, with an eight-day run (offering $70,000 in prizes), the Royal has constantly mushroomed. Although even then, in an eruption of local pride, the *Toronto Star* acclaimed it: "The Best And Biggest . . . In All The World."

Part of the hoopla was doubtless due to "California Frank's Mammoth Western Circus" in the Coliseum, and Canada's first (and *only*, as far as we know) "Walking Goldfish" competition. But most of the excitement was due to the horses.

Both the Royal Canadian Dragoons and the Governor General's Horse Guards performed in the Jumping Stake. Farmers packed the outdoor stables to ogle the nation's finest specimens of Belgians, Clydesdales, Hunters and Jumpers. There were even competitions between the different classes of delivery horses — used by bakers, butchers, ice plants, express companies and dairies — that cluttered each city street. In fact, one Toronto dairy on Spadina Crescent claimed, "We serve more homes than any dairy in the British Empire."

For those were the days when the horse was supreme. When the first fair opened, Canada had fewer than 1 million citizens — and 3,624,300 horses. Even the smallest farm owned draft or riding horses, since tractors were another 20 years away and motor vehicles were still a novelty. Standardbreeds and hackneys dotted every country lane and cobbled street.

Yet every era ends. With the passing of the buggy and horse-drawn plows, even a deputy agriculture minister asked if the Royal would survive in an industrialized society. Both the horse set and the farmers were aghast at his gall. And over the decades, except during wars, the fair reopened each November.

Still, by the eighties, attendance was plunging and exhibitors were dropping out. In 1988, the day of reckoning nearly arrived when the fair sank a whopping $500,000 into the red. It was mainly due to the management skills of the directors Galen Weston, Byron Beeler, Brian Conacher and John Craig Eaton (whose mother was one of the fair's first great forces) that the Royal made an abrupt and dramatic turnaround. By 1991, the revitalized fair saw attendance soar by 90,000 from the year before, while turning a neat $350,000 profit.

Ironically, with farm families down to 3 per cent of today's population, the Royal Winter Fair has never been healthier. California Frank's Mammoth Western Circus may have been superseded by the Superdogs, and the Midway replaced by the Petting Farm, but the stallions still leap in the Coliseum, tiny calves still wobble in the Cow Palace and the Royal Winter Fair is still the biggest in the world.

The 40,000 school kids who stampede its doors couldn't ask for more.

1 The Birth of the Royal

CANADA HAS HELD FAIRS AS LONG AS IT'S HAD farmers. After clearing their land and erecting log houses, the early settlers built schools for their children and churches for their souls. Then they organized country fairs.

In 1765, the first fair in Canada was held in Windsor, Nova Scotia. Niagara-on-the-Lake had the first in Upper Canada seven years later. Then, through the 19th century, the trend swept the country. Several were running in the Northwest Territories before Alberta and Saskatchewan became provinces. Yet with them all, as fair historian W. P. Watson notes, the emphasis "was on quantity rather than quality." The hugest ox and fattest pumpkin were considered best.

In the early 1800s, farmers began importing pure-bred livestock from Britain — the source of most homegrown breeds today. As the pedigreed herds developed, the new Canadian "breeders" began arguing about which breeds were best — and which were the best animals in each breed.

The first question is still being debated. The second was settled in

1850, when breeding standards were established and animals were divided into age categories. And so today's system of competition and expert judging was introduced.

By 1910, every province was hosting agricultural fairs. Besides the Ontario Provincial Winter Fair, held annually in Guelph, the Ottawa Winter Fair was also setting livestock standards. Yet Canada still lacked a national fair to draw exhibits from across the country. And Will A. Dryden, an Ontario Shorthorn breeder, thought it was time.

After he formed a committee with three other ardent Ontarians — Guelph professor George Day, D. C. Flatt and George Pepper — the men met with government agriculture ministers. A national fair, they insisted, should feature competitive classes for all breeds of Canadian livestock and types of produce. But recognizing that such exhibits, no matter how prizeworthy, would hardly dazzle the general public, they also suggested including a horse show.

And so, with government backing, the National Live Stock and Dairy Show was held in Toronto in 1913. It was intended to be an annual event, but there was one hitch to the timing: within a year the world was at war.

It wasn't until late in 1917, when an Allied victory seemed assured, that the four men met again to revive their vision. After convening a meeting in Hamilton of leading livestock breeders, a seven-man committee (comprising Clydesdale, Ayrshire, Shorthorn, swine, sheep and poultry breeders) was appointed. The following February, members of every breed association met in Toronto's Prince George Hotel. After electing Will Dryden chairman and George Day secretary, the 48 delegates began planning a national fair — large enough to accommodate 10,000 poultry, 1,000 sheep, 1,000 hogs, 800 beef cattle, 700 horses and 600 dairy cattle.

Any interested city was given the chance to bid as a permanent site. When all proposals were in, a committee boiled them down to three: Toronto, London and Hamilton. Each city was allowed to argue its merits. At the same time, Canadian National Exhibition officials were approached about using their livestock buildings — but *only*, as Will Dryden stressed, if the proposed fair was controlled by the agricultural and breed associations, and *not* the CNE.

At the end of all discussions, the committee announced its decision. The fair would be held in Toronto. But when several original members furiously objected, the long-suffering Will Dryden called a final meeting, in October 1919, to settle the issue. Since London's claim had meanwhile been rejected, the choice was narrowed to Toronto and Hamilton.

Toronto offered to erect a $1 million building in the CNE grounds — and provide free power, water, light and heat. In return, the fair would repay the city $40,000 annually for ten years, plus $10,000 a year rental for the same period.

Hamilton pledged a free site, plus $20,000 annually for 20 years to help defray building costs (estimated at $800,000), as well as all utility expenses. Or, if the fair preferred, a flat grant of $250,000.

Before the final vote, supporters of each city argued eloquently. So convincing were both camps, in fact, that when ballots were counted, Toronto and Hamilton had 18 votes each. The tie-breaking decision fell to Chairman Will Dryden, the Shorthorn breeder whose dream the fair had initially been. He cast the deciding ballot for Toronto.

The CNE location was now set, along with Toronto's promise of a permanent building on the site. Yet the reality of a national event still depended on the fair getting $50,000 in federal and provincial grants for prize money — plus raising $50,000 annually on its own.

Both Queen's Park and Ottawa responded smartly — each agreeing to annual grants of $25,000 for a decade. Then, after the breed associations offered to supplement cash prizes for their species, private donations started pouring in. The Union Stock Yards pledged $2,500 annually for ten years, while eight companies, such as Swift Canadian and Eaton's, promised $1,000 a year for the same duration.

The organizers were elated. At a November meeting in 1919, the committee incorporated itself under the name "The Agricultural Winter Fair Association of Canada" — with C. F. Bailey of Ontario's Agriculture department as managing director. But even before Bailey took over, certain members were sniping at the fair's plodding title. The word *Royal*, they insisted, should be included. Why? No reason, except sheer colonial snobbishness. Yet after King George V gave his blessing, *Royal* was prefixed to the title.

Such vital decisions took time. It wasn't until May 1921 that construction of the 7,500-seat Coliseum, with adjacent annexes, finally began. Expecting the work to take six months, the organizers advertised the Royal's grand opening for late November. But by October, after exhibit space had been rented and programs printed, it was obvious the arena would not be ready.

The fair's cancellation was a serious blow to the Royal — both emotionally and financially. Even with government and private grants, its agreement to pay the city $40,000 a year, plus $10,000 rent, was suddenly deemed impossible. One group of directors, led by George Pepper, argued that the Royal should be vested in the CNE. Another, under C. F. Bailey, decried such action as a betrayal of the association's principles. So fiery grew the debate that both men resigned. Finally, the old pact with the city was scrapped, and the fair agreed to pay Toronto all annual profits over $10,000.

The next opening date was set for November 22, 1922. And this time the Royal was ready. Its blaring ads promised that the fair would "Reveal the Marvels of a Winter Exposition Unprecedented in the History of Any Nation in the World!"

Along with "California Frank's Mammoth Western Circus" and a "Horse Show Extraordinary," fairgoers were offered a "Modern Indoor Midway" and "The World's Greatest Display of Livestock." The public gobbled the bait.

The CNE grounds were soon busier than Union Station. More than 200 people arrived on a special train from P.E.I., while the Grand Trunk began routing three trains a day to the fair from Hamilton, Brantford, Woodstock and London. Meanwhile, 225 freight cars of livestock rolled in from every province and nine American states. The 17,000 entries included 2,500 cattle, 1,850 horses, 900 sheep, 700 swine and 9,100 poultry — all competing for $70,000 in prize money.

At 8:00 a.m. on November 22 — a cold and cloudy Wednesday

morning — the box offices opened. The first 25-cent admission tickets were bought by Parry Sound farmer E. A. Haines and his wife. And 22,000 others joined the lineups that opening day. Reserved seats for the horse show (50 cents to $1.25) sold briskly, while the rich forked out 50 dollars for a six-seat box.

It was, indeed, the event of the year. They poured in by streetcar, Grank Trunk and CPR, as well as shiny Packards, Maxwells and Durants. At the horse show, a full-dress orchestra played. There were silk hats, evening gowns, ostrich plumes and, said the *Star*, "pails and pails of jewels."

There was just one recorded mishap. Two young girls placed bets on the outcome of the Hunter class, handing their money to a fellow spectator — who disappeared the instant their horse won. To compensate for their loss, officials let them into the California Frank's Mammoth Western Circus free of charge. (Though Frank took top billing, his wife — who'd performed with Buffalo Bill before achieving fame at Coney Island by diving her stallion off a 40-foot platform — was really the star.)

Before the week ended, more than 150,000 excited souls had passed up seeing Valentino at Shea's Hippodrome or Rachmaninoff at Massey Hall to catch the Royal's marvels. And the superlatives poured in. The *Toronto Star* hailed the fair: "The Best and Biggest of Its Kind on the Continent If Not in All the World."

The Toronto *Globe* called it simply: "Greatest on Earth." Sounds as if they liked it.

At the time of the initial fair in 1913, more than half of Canada's purebred livestock breeders were in Ontario. Most western farmers (except for Alberta ranchers) were content to grow wheat. And during World War I they prospered. But within a decade, wheat prices plunged from record levels to less than 50 cents a bushel. As the grain markets disappeared, desperate farmers switched to livestock. Soon the western breeders of purebred cattle, sheep and swine outnumbered those in Ontario.

When they started railroading their livestock to the Royal, Alberta and Saskatchewan began donating annual grants of up to $5,000 — as did Quebec, New Brunswick, Nova Scotia and Prince Edward Island in the east. Soon all provincial governments, except Ontario, made agreements with Ottawa to reimburse fair exhibitors for expenses and transportation costs. In return, the provinces had the right to choose their own potential prize-winning entrants. By establishing national standards, the best Canadian livestock was brought together.

There was just one problem: the stalls they were placed in were woefully inadequate. While poultry and farm products were displayed in the annexes, the livestock was housed in 30 separate structures beside the train tracks north of the arena. Although adequate during the summertime CNE, in November the frame stables were cold and wet, with a bog of mud outside. Cattle spent hours standing in water; stablehands caught pneumonia. Such conditions couldn't continue.

In 1926, the Canadian and Ontario governments agreed to share the $1.4 million cost of constructing new buildings to house 2,000 cattle, 1,200 sheep and 1,000 swine. Work was completed in time for the 1927 show. Yet horses were held in the makeshift stables until 1931, when the three-acre, 1,200-stall Horse Palace with its ornate limestone façade and bas-relief friezes, was constructed. At an estimated cost of $900,000, to be shared between the governments of Canada, Ontario and Toronto, the project was a means to ease unemployment during the Depression.

Still, though more than half of the building costs were paid by Ottawa and Queen's Park, the City of Toronto owned the fair. In return for the gift, the city promised to maintain the premises for the use of both the Royal and CNE. Which, until 1939, it did. Then, that September, the world again roared into war.

The Coliseum and its adjoining buildings were among the first to be requisitioned for military barracks. For the next six years, thousands of troops were stationed there during the fair's cancellation. Although the Union Stock Yards offered their facilities for special livestock exhibits and sales from 1939 to 1941, the shows were stopped when meat rationing began.

No livestock event could compare to the Royal, however. And when the first postwar fair opened in 1946, hundreds were turned away from the jam-packed site. The buildings were simply getting too small for the increasing crowds and number of livestock entries. In 1949, Ontario offered a $1-million grant to enlarge the facilities, if Canada and Toronto matched the offer. While Toronto's mayor agreed at once, it took Ottawa three years to join in.

Again, a building committee was set up. And once again, its members disagreed. Some wanted the money spent on bigger stabling space; others on judging facilities. The CNE also wanted in on decisions. Finally, the entire committee was sacked, and a new one formed as a joint Winter Fair–CNE endeavour. It failed as well.

It wasn't until 1959 that Ontario's Minister of Agriculture said its grant would be dropped if action wasn't taken instantly. So, of course, another committee was formed. A year went by. At the end of the 1960 fair, construction finally began. The old sheep building was replaced by a two-storey sheep and swine structure, which would hold 2,000, animals. The Cattle Palace got additional stalls for 400 head, and indoor pens were added for steers.

In 1961, a new milk house was built and equipped. In the arena, all steel posts except four in the corners were removed, and the reserved-seat area was redesigned. A new façade and huge lobby with a fountain were added to the Coliseum. And a new herdsmen's dormitory was erected.

It had taken 12 years since funds were first offered, but at last the Royal had world-class accommodations. Three years later, it took another leap forward by holding its first Sunday horse show — although other fair activities were shut down on that day. It was not until 1968 that the Royal Winter Fair was allowed to run full blast on the Sabbath. Toronto the Good and its agricultural showpiece on the waterfront had finally come of age.

2 *A Day at the Fair*

IT'S SATURDAY AT 10:00 A.M. A GOLDEN, CRISP November morning. In the sprawling grounds of the CNE, the trees stand bare.

At the eastern tram loop near the Princes' Gates, they pour off the double-linked streetcars and stream through the doors. Toddlers on their daddy's shoulders. Moppets clutching grandma's hand. Queues of giggling students trailing teachers. A tidal wave of tykes.

Although 10,000 animals are trucked to the fair each year, school kids outnumber them four to one. For each star steer or stallion on display, there's a quartet of awed young fans. Most have never in their lives seen a llama — to say nothing of a duck that plays piano. But they're on display at the Royal.

The duck performs at the Petting Farm in the eastern wing of the fair's six massive buildings — where most of the kids head first. She stands in a case beside a tiny yellow piano with a mini lamp on top, staring back through the glass at the mob of little faces. But slip a loonie in a slot and *presto!* As honky-tonk music bursts from the box, the daffy duck spins around and starts pecking the ivories, jerking her bill up and down the keyboard. The children gape and giggle.

Behind them, a big man with a ponytail watches the performance. "Darn it," he mutters, "she's supposed to turn the lamp on. That's a new duck I'm teaching, but she hasn't learned the trick yet." Just then the concerto ends, and the duck swerves around to gobble pellets plunking into a cup. "That's the feeding machine," says the big guy. "It's the first thing I teach them."

The next case over holds a chicken called "Bird Brain," who plays tic-tac-toe on an electronic board. Beside it is another hen that bops baby basketballs into a net. "They all work for birdseed," quips the ponytailed man, Ken Lasalle. He keeps 300 small animals on his 40-acre Oshawa farm. And he's brought half his menagerie (in three spacious semis) to the Petting Farm enclosures.

Rows of fenced-in pens along the walls hold a mother and baby zebu from India, two Vietnamese pot-bellied pigs, Samantha the zebra, a Montanan Spotted Ass, an Australian emu ("Some of these guys," says Ken, "cost up to $20,000"), a Peruvian llama called "Doc Buzz," a two-humped Bactrian camel, a Hunchbacked Zebu bull ("The Brahmin bull's forefather") and an amazing Four-Horned Scottish Jacob sheep, with each horn aiming flat in four directions.

In the centre of the room sits a huge rail-fence corral stocked with dozens of baby goats, sheep, pigs, llamas, donkeys — and euphoric human urchins. The kids hug, kiss and feed the fuzzy creatures, struggle to lift them and invariably pipe up, "Hey Dad, can we take it home?"

Lasalle smiles fondly at the children's delight. "We've got the only petting zoo I know of where kids can go into the pens," he says. "The only problem is, they never want to come *out*."

Next to the farm, in the fair's sheep and goat section, five snow-white ewes are tethered to posts as their owners swiftly shear them, briskly brush them and mist their coats with an aerosol spray. They're then completely shrouded in canvas capes and hoods (to keep them squeaky clean) and led back to their pens. Staring through their balaclava eye slits, the docile creatures seem suddenly transformed into sinister Disney terrorists.

In the adjoining Fraser Pavilion, hundreds of farmers jam the main floor bleachers and tiered balconies above, all overlooking a central show ring inexplicably packed with garish pea-green shavings and skirted by a white picket fence. A cattle auction is just beginning, and the farmers — sporting bill caps, Stetsons and thin-brimmed felt fedoras — are flipping through glossy programs titled "Sale of Stars."

At the west side of the room stands a small raised stage bordered with potted yellow mums and ringed with ropes. Nervous Holsteins are led up a ramp to the tiny bovine auction block — where they stand, mooing loudly, as the auctioneer spurts his non-stop spiel.

He's a little man in a giant white Stetson, clenching a gavel in his fist. Mickey Rooney playing Spencer Tracy. Though the buyers seem to understand his babble, laymen would kill for a translator. Spitting words like tracer bullets, he's Porky Pig on speed.

He only sounds coherent when, after a bid, he stutters, "I got twenny . . . twenny . . . twenny . . . need twenny-one . . . twenny-one . . ." Followed by a further incomprehensible spew.

Still, every few seconds the price jumps. Men in suits stand around the arena, eyes darting — shouting out only when a secretive bid is made. These guys are *good*! You never notice a single nod or gesture in the crowd.

A tiny calf is led up the ramp, so young its stilt legs wobble. Onstage, it looks dazed. "I got nine . . . nine . . . nine . . . need ten," spits the Stetson. "Ten . . . ten . . . ten . . . *Ten*! Got it! Need eleven . . ." It takes less than a minute of further babbling. The calf goes for fifteen grand.

A stolid Holstein stares at the crowd as the bidding soars. Then, raising its tail, delivers its own overt opinion of the auction. "Forty-five . . . forty-five . . . forty-five," the auctioneer brays into his mike. "Need forty-six." He speaks in tongues for twenty seconds more, then whacks his gavel. "*Sold*! Forty-five thousand. She goes to Ireland."

In the Swine Barn next door, scores of pigs with green numbers stencilled on their bristled backs loll in straw-littered stalls. They're the *cleanest* swine we've ever seen. "But," insists one of the owners, "pigs *are* clean. The trouble is that they don't sweat, so they have to roll around in mud or water to cool off."

Surrounded by wide-eyed waifs, she's standing beside a display pen. In it, a 400-pound sow stretches out asleep — while ten week-old piglets scramble towards her teats.

"Look at *that*!" cries the woman, as one piglet wiggles past its mother to an opposite corner. "See what I mean? They always sleep on one side of the pen, and poop in the other." She beams devotedly. "That's all they do. Poop and sleep. But mostly eat. In the twenty-one days till they're weaned, they gain a pound a day."

Then, pointing at the sow, she says, "You notice that she's got fourteen teats. But four have dried up because they're not being used. That's because all ten piglets choose their own personal teat. And once they select one, the rascals never share it. They make their choices fast.

"They're born, you know, with their eyes wide open."

Through an entrance to the west stretches the famous Cow Palace, a massive space roofed by towering tiered skylights. Green-and-white bunting, plus a gigantic Canadian flag, dangle from the overhead steel spans. The barn is redolent with the smells of fresh straw and manure, and resounds with the mournful tuba groans of cattle. Hundreds of dairy cows. Jerseys. Holsteins. Ayrshires. Guernseys. Brown Swiss. Milking Shorthorns. All in separate stalls, abutting both sides of concrete aisles.

In one aisle, a young man named Bernie clips a docile Jersey. Bernie, who works for a major Richmond Hill farm, is one of eight employees (including the night shift) at the fair. He's called a "fitter," he explains, since the hour-long haircut makes a cow "fit to show."

Along with ten cattle, the crew has brought a microwave, refrigerator, milker and portable vacuum pump, as well as seven bales of straw (to mix with seven bags of pine shavings). "Coming to the Royal isn't cheap," he admits. "After buying feed and paying the help, it costs about five thousand bucks. Some farmers consider the fair a marketing tool. Others just come for the annual reunion."

But no one considers it a vacation. Bernie says they feed and water their animals six times a day, clip and shampoo each one before a show, and milk them up to three times daily. "The first week of the Royal," he says, "features goats and beef cattle. The second is for swine, sheep and dairy cattle.

"And during that final week," he enthuses, "about 42,000 litres of milk are produced. That's 6,000 gallons a *day*. And every drop is weighed, tested and recorded for its butterfat content before being trucked off to dairies."

At the fair, Bernie himself works 19 hours daily. "I hit the sack about eleven each night, and get up again at four." And where does he sleep? "Right here," he grins, pointing at the straw in the stall beside him. "Just crawl into my sleeping bag."

He pats the back of the Jersey he's just clipped. "Dreamgirl here gets mad if I don't sleep beside her."

In a corridor aligning the Cow Palace, rows of glassed-in showcases are crammed with seeds. The displays are part of the Field Crops Show and represent the fair's prize-winning specimens of seed corn, timothy, hay, soybeans, barley and wheat.

Though hardly a huge draw for the average urbanite, knots of farmers hunch over the cases like devout Dead Sea scholars. Many scribble the listed statistics into note pads.

One sun-weathered old man in a shiny grey suit and plaid lumberjack shirt peers into a display of dried tobacco leafs. "Hey, Millie," he calls, "come see *this*!"

Millie, who's trying to control two hyper grandkids, hurries over, darts a swift glance and nods curtly — as the kids scamper off down the corridor. "Awright, Howard," she hisses, "that *does* it! I *swear*!"

When Millie catches up to the kids, they're ogling a team of lifesized plastic horses hitched to an authentic antique threshing machine — while gleefully punching a red button that rotates its wooden blades. With each jab, the bright red thresher coughs alive and starts twirling. When the button's released, it creaks to a stop like an asthmatic windmill. The kids are ecstatic. Poke, spin, stop. Squeals of delight. Jab-jab-jab.

Millie seems on the verge of tears. "I *swear*!," she mutters, tugging them away. But the poor lady's salvation lies right across the aisle. Two smiling women in scarlet shirts are demonstrating a 100-year-old wheat grinder and beckon the imps over. "Hey, kids," calls one, "come, give us a hand."

And so, as the little girl dribbles fistfuls of wheat into the relic, her brother cranks the handle. Flour spills into a tray beneath. The kids smile shyly. Even Millie seems pleased.

"Now *wait!*" says one of the women, passing each sibling a sieve. "You've still got to sift the flour from the bran."

By the time their granddad shuffles up, the kids are sifting up a storm.

"Well now," beams Howard, "isn't that *some*thin'? We'll make farmers of these two yet, eh, Millie?"

Millie's glare could roast raw tobacco.

At the south end of the corridor, a banner stretching above a door reads: "International Trade Centre." It's one of the few rooms at the fair that the public can't enter. So what goes on inside?

Well, as the Royal's press book says, "It's a meeting place for exhibitors, buyers and exporters to discuss and promote sales . . . of purebred breeding stock, semen and embryos."

A handprinted sign beside the door says: "Messy Boots Allowed." Sounds intriguing. So we enter.

A young lady in tailored jacket and skirt sits at a desk inside. In the room behind her are clusters of round tables and tiny writing desks. Half of the tables are occupied, mostly with men in suits. In a corner at the rear stands a giant soft-drink machine.

The receptionist says the lounge is reserved for Canadian breeders and foreign buyers, "to meet and make deals." She smiles. "We try to make it as efficient as possible . . . by providing phones, office equipment, faxes, video sets, import/export information and refreshment service."

And how many foreign visitors come to the centre? "Well, there's more than 800 registered. But dozens just drop in."

Where do they come from? "Gosh, you name it. We've got buyers here from 35 different countries. But most are from Japan. And they've paid up to $500,000 for a single Holstein."

Just then a tall, flaxen-haired man in a houndstooth jacket comes to the desk. "Excuse me," he says, in an accent weighted with umlauts. "Do you please speak German?"

"No, sorry," says the woman, rising. "But I think there's somebody here who does." She sweeps into the room, swiftly returning with one of the buyers. "This," she says, "is Mister Graf from Zurich." The men converse, shake hands and head inside.

The centre doesn't usually provide translators, she says, "since most groups bring their own. But this year, the Ontario Hereford Association invited a large Czech contingent over. So we did provide one for them."

One final question. "If someone's interested in discussing semen or embryos, who should they talk to?"

She jabs a finger at the door. "Just go out *there!*"

"*Sorry*," we stammer, "that wasn't meant to sound offensive."

"Yes, I know," she nods. "In the corridor outside are a group of people in different booths who'll be happy to discuss *any* breeding procedures with you." She smiles and says, "Good luck."

And sure enough, there *is* a row of horse and cattle breeding booths outside — with agents behind the counters.

We stop at the first. It's run by Alta Genetics Inc. out of Balzac, Alberta. The man at the stand says Alta has dealt in cattle semen and embryos for 25 years. "And today," he adds proudly, "we export them to 47 countries — from Australia and China to . . . well, Egypt and Zimbabwe. Our marketing agent, Semex Canada, is the biggest exporter of dairy cattle semen in the world."

When we say we're impressed, he smiles.

The frozen semen and embryos are transported in liquid nitrogen containers, he explains, while live cattle usually travel in the cargo holds of commercial airlines. "But we also have offices now in Argentina, Brazil and Mexico. And last year alone we trained professionals in seven countries from Chile to South Africa."

And what exactly are they trained *in*? The salesman beams, as we scribble notes. "Oh, a vast variety of scientific techniques. Reproductive physiology. Biotechnology. Embryo recovery, evaluation and freezing. Fetal sexing by ultrasound. Alternative embryo freezing techniques. Embryo sexing. We're renowned for quarantining and exporting elite genetics in . . . "

Whoa! Impressive stuff no doubt, but daunting! Just then we spot Millie and her gang passing by. "Oops," we say, "there's the family. Better catch up." We grab a brochure.

The Alta man looks puzzled. "You're not, ah . . . in the breeding business, are you?"

"Not exactly," we say. "But we're definitely not against it."

Above the Petting Farm on the second floor, there's a long queue stretching past the rabbit hutches. The kids and parents are clutching free passes for the Superdogs show, the fair's top attraction, with up to five performances daily. Along both sides of a massive room, tiers of benches seat 1,000, while scores more squat on the floor. (Many of the kids are wearing cutout cardboard crowns blaring the words "Medieval Times," a huge amphitheatre in the CNE grounds where armoured knights on horseback joust and duel within an arena as surrounding spectators feast on chicken. Having just witnessed the charging steeds, they've rushed across to the Royal to see the dogs.)

A man with silvering hair strides into the room. He's wearing a white jacket, purple slacks and Hawaiian shirt, and clutches a cordless mike. The Superdogs, he announces, "is an all-Canadian show. It entertains one and a half million people each year — more than any production on Broadway."

There are 20 dogs in each performance, ranging from Newfoundlands, German Shepherds, Border collies and Rottweilers ("who've been around for 2,200 years"), to keeshonds, Jack Russell terriers, toy and miniature poodles. ("Poodles are German, *not* French. And they're one of the few breeds with no traces of colour — always solid.")

As he talks, attendants set up two rows of hoops and hurdles down either side of the floor — and suddenly the dogs scramble out of an entrance at the end. Led on leashes by their (mostly female) owners, they split into two relay teams.

Each team, we're told, must dash across the obstacles and poke a machine at the end — which pops out a tennis ball. Snapping up the missile in mid-air, the dogs should then streak back through the obstacle course. Spectators on either side are cajoled to cheer on their team. And may the best mutts win.

Taped music blares. A whistle screeches. And the kids go wild as the absurdly mismatched pooches take off. A tiny Pomeranian down one side, a hulking Newfie down the other. A Border collie and poodle next. Down the course they dash, arch and soar. Great galloping lunges by the big guys; tiny piston-pumping strides by the shrimps. Yet every one a die-hard Olympian going for the gold.

It's truly a riot. They're funnier than Chaplin. The crowd is screaming in seconds. A keeshond misses the ball. It bounces. The keeshond soars to snatch it. The ball caroms off keeshond's snout and bounces into the bleachers. The keeshond streaks behind it, leaping across laps. Kids go insane. A collie leaps so high it smacks the top of hoop. Hoop topples. The collie back-flips, stares around, saunters home in disgust. A yipping poodle ignores the course completely, scampers beside it, snaps ball in its teeth, then simply *sits*.

The teams then compete in high jumps, bars being added after each heat. The Newfie draws applause for hoisting such bulk so high. The poodles get cheers for making it at all. But a mottled black-and-white Border collie named Hampton gets cheers for his sheer showbiz intensity. When Hampton leaps, all four limbs stretch in a ruler-straight line. He looks (without the silly blue body suit) exactly like the soaring Man of Steel.

After each successful series of hurdles, Hampton snatches a cap from his owner. Then he chomps it, snarling, shaking it like a rat, all the way back to the starting line. Hampton's pale blue eyes, up close, seem insane. But then again, Jonathan Winters — a fellow master of madcap timing — frequently checked himself in. Fortunately, when the Superdogs play Musical Chairs (on wrestling mats), Hampton is chosen to star. To the skirl of Scottish bagpipes, ten dogs parade around nine mats. When the music suddenly stops, they're supposed to squat on the nearest pad, with the odd mutt out disqualified. Then, before the music restarts, Hampton's job is to zero across the room and snatch away a mat.

But usually when the music ends, at least two pooches pounce on the same pad — leaving the others unoccupied. Hampton, however, hates crowded mats. He always zooms in on the one *most* crowded, snarling it off across the floor, its baffled occupants spinning like pinballs.

In the next show, the Superdogs play baseball.

And the amazing thing about them is that few of the hijinks are rehearsed. Hampton, for instance, wasn't *taught* to choose a crowded mat. He does it because he *wants* to. As Herb Williams, the show's announcer and owner, says, "No two shows are ever the same. We just never know what the dogs will do."

The dogs, in fact, aren't really trained performers at all. Nor are they the same actors in every show. "Across Canada, we have about five hundred owners or trainers with a thousand dogs that we use at different times," says Williams. "When we're putting a show together for a certain city, and we do about forty cities a year, we contact the people there.

"We know we need so many small dogs, big dogs, and dogs with certain attributes — like the Newfie who jumps so high. And we make our selection from the choices at hand. Today, in fact, the show's so popular that owners come to *us*. I devise the routines and choreography, teach the owners what we want, and then *they* train their dogs."

Though twenty pooches may compete in one show, Williams says, more than fifty might be rotated through a day. During the 12-day run of the Royal, including six special shows in the Coliseum, the dogs lure crowds of 85,000. In return, the show receives $8,000 a day from different sponsors.

But the Superdogs shows also travel on the road for 200 days a year. Besides a permanent crew of six, including soundmen and drivers — who haul $100,000 worth of props and equipment in two trucks and a 50-foot tractor trailer — there is also a "cast" of forty owners and their charges, who travel by air or private van. In return for their efforts they get hotel rooms, travel costs, all expenses and $40 a day. They're obviously not in it for top showbiz bucks.

Overseeing it all is the energetic Williams, who, before devising Superdogs in 1980, was a schoolteacher, an ad man and a publicist for General Motors. Today, he sometimes has three shows on the road simultaneously, including a two-and-a-half-hour theatre production. In the few weeks he's home each year, Herb raises Chows.

On the second floor of the east annex, a few dozen fairgoers (mostly women) perch attentively on benches in front of the Kraft General Foods Canada stage. The stage is designed as a kitchen, and a demonstrator stands at a table beating eggs in a bowl. Above her head hangs a large tilted mirror so the audience can observe her culinary magic.

"Now," she says, "in a separate bowl, you combine the flour, cocoa, baking powder and salt." Which she does. "Then gently stir the flour mixture into the egg mixture until blended. See? When that's done, you fold in melted chocolate and . . . "

The chin of a man in the back row slumps slowly to his chest. His eyes are shut and he looks at peace. But suddenly his head jerks up when the woman beside him jabs his ribs. For a full sixty seconds he tries valiantly to concentrate.

". . . And now, you spread the batter in a greased eight-inch-square baking pan before . . . "

The chin again begins to sag. This time, his wife simply darts a killer glance. By the time the demonstration ends, he's snoring softly. The lady on stage invites everyone to "come back at 3:00 p.m., when all the prize-winning pickles, jams, jellies, honey and maple products will be auctioned off for charity."

The women burst into applause. The head snaps back up. The guy looks around and claps like crazy.

A few yards from the stage, busy stalls are hawking everything from hams and hot dogs to honey in the Country Food Fair. A man in a booth is selling venison sticks (which look like long salamis), and he's offering sliced samples on a plate. "Oh yes," he assures one taster, "venison's far leaner than beef."

Business is brisk when he offers special deals. The vendor busily stuffs plastic bags with his venison sticks, which he takes from a glass display case. We even buy a couple ourselves.

It isn't until two days later that we spot the printed label on our purchase. In bold letters running down its length, it boasts: "All-Beef Salami Stick."

Well, Bambi at least would be pleased.

Across the room from the venison vendor, a gaggle of kids and their keepers are clustered around a flat, glass-sided beehive frame. Of the 1,500 bees swarming through the comb inside, the tots are trying to pick out the queen. "I'll give you a hint," says the beekeeper. "She's got a green spot on her back."

After a moment of intense concentration, a youngster yelps, "I *see* her!" And sure enough, there's the monarch with the green spot, ringed by a protective guard of workers. The other kids say, "Wow!"

The beekeeper smiles. "Good for you, son. But this is only an observation frame. A regular hive holds almost *eighty thousand* bees." The kids say, "*Wow!*"

The beekeeper says that most queens live three years or so. "The one you see here — after mating with about a dozen drones — laid all the eggs in the hive. And each egg becomes a bee."

"You mean," chirps the keen-eyed kid, "she's not only the queen, but their *mum*?"

"You *got* it," the beekeeper winks. "She's the mother of them all."

"WOW!"

On a table nearby squats a stupendous, if sadly sagging, pumpkin — a winner in the fair's Giant Vegetable Competition. A card beside it says the orange monster weighs 658.4 pounds. Arrayed around it are other massive jack-o'-lanterns-in-waiting. Squashes the size of baby Goodyear blimps. An assortment of obscenely bulbous cabbages, onions and potatoes. And even garlic bulbs as big as baseballs.

As fairgoers gawk, one prize-winning grower (in brushcut and beard) is extolling garlic's medicinal virtues. "The only thing wrong with garlic is the smell," he says. "Otherwise, it's certainly nothing to sniff at." He grins at the ripple of giggles. "But seriously, there are 600 varieties of garlic around the world. And scientists have found that the chemicals in them may reduce the risk of heart disease. What's more, they can also lower blood pressure by up to 10 per cent. While a single clove of garlic can reduce blood cholesterol by 9 to 12 per cent."

"In that case," cracks a young woman in braids, "I guess it's worth the stink."

"Sure is," laughs the grower. "But milk and cranberry juice, or raw celery and parsley, will kill the odour. And if you work up to a clove a day, and your system gets used to it, there's no tell-tale breath at all.

"You see, garlic gets its odour from an amino acid called allicin, which forms when cloves are cut or crushed. The highest amounts of allicin come from Chinese or Argentine garlic. The bulbs we grow in North America are only half as potent. Yet allicin doesn't last very long, and it's quickly killed by cooking."

"So it's best to eat garlic raw?" asks the braided new devotee.

"Absolutely," says the grower, grinning. "Unless, of course, you're an elevator operator."

The adults chuckle at his spiel. But fidgety offspring keep tugging them off to the Poultry Show — where raucous roosters, hens, pigeons, turkeys and geese (including a few Canadian honkers who missed the last flight south) strut around in wire cages. A sign behind the chicken coops says, "The average bird in a flock lays 300 eggs in 52 weeks."

There's suddenly a whoop from the moppets. Clomping towards them down the aisle is a silly-looking guy, with a huge red nose and rainbow wig. His black lace-up boots are the size of snowshoes, and he's leading an invisible dog on a bobbing leash. To his pals, he's Peter Thompson. To the kids, he's Puffy the Clown.

The kids squeal at Puffy to show them some magic tricks. Puffy says he would, but he has to walk his pet. "Whaaaaat pet?" they shriek. "There's nothing *there!*" Yet, three feet in front of him, the leash still wiggles.

"C'*mawn*, Puffy, show us a trick," they persist. Puffy finally gives in. He takes a wooden box from under his arm, sets it down and slides open an empty drawer. Then he unleashes his invisible pooch, places it in the drawer and shuts it. For the next ten minutes Puffy pulls quarters out of toddlers' ears, makes cards disappear and turns purple balloons into toy giraffes.

The kids scream for more, but Puffy has to go. "Bye, bye," he says.

"Gotta take my pet for another walk." He reaches down and opens the drawer. Inside sits a big white rabbit.

The kids out-whoop the roosters.

In a small judging ring downstairs, a line of Suffolk sheep breeders in white shirts and dark slacks is circling the sawdust. On their heads they wear white cardboard coronets stamped with individual numbers. And each is leading a sheep. In the centre, a judge in grey suit and fedora peers intently as they pass.

A farmer's wife from Orillia sits in the tiny bleachers and describes the procedures to a friend. "That judge there, he's a Suffolk breeder himself, from Scotland. They often use judges from Europe and the States."

"What's he checking for?"

"Well, first of all, Suffolks must have pure white bodies, and totally black heads, ears and legs. That's essential."

In the ring, the sheep are being lined up side by side, their white bottoms facing the bleachers. The friend suddenly gasps. "My gawd, what happened to their *tails*?"

"Oh," laughs the woman. "They cut them off when they're lambs."

"Whatever *for*?"

"Well, I guess it makes 'em look neater." She shrugs. "Anyway, they don't really need them."

"Uuuugh!"

"Now look how the judge is staring in their mouths. He's checking to see that their bottom teeth meet their top jaw evenly."

"You mean their top *teeth*?"

"No, their *jaw*! They don't have any top teeth in the front."

"Gaaawd. And what's he doing *now*? Grabbing them all like that?"

"Well, he's checking them for length and weight, and the depth of their chest wall. And their meat, of course. Suffolks are a valuable meat breed. They can sell for up to $1,000."

Just then, a loudspeaker announces the winners. The small crowd claps. The judge hands the First Prize ribbon to a beaming young brunette. "That's worth sixty dollars," says the woman from Orillia.

"You're kidding? A measly sixty bucks for a thousand-dollar sheep?"

"Hey! To get a Royal Winter Fair ribbon for breeding is something. They don't enter for the money. It's *prestige*!"

The women stand. "Well, I'll tell ya one thing, Sal," says the friend. "Don't give *me* a sheep with no tail."

In the east annex, near the main entrance, sits the celebrated Winter Garden Show. The first thing you see are the tulips.

A sea of yellow, red, mauve and white tulips. Fifteen *thousand* tulips, in fact. And because every bulb was replaced halfway through the show, there've been 30,000 tulips in all.

Since the fair is celebrating the 400th anniversary of the tulip's arrival in the Netherlands (from Turkey), the Winter Garden looks like a tiny hunk of Holland. A 100-foot cobblestoned Dutch canal runs across the room, with tulips bursting along each bank (though they actually arrived from New Zealand).

There's a short streetscape of Amsterdam, with building façades and pocket gardens, a Flemish courtyard, a working greenhouse and a "Master Gardener" in an information booth. A stone and iron footbridge crossing the canal leads to a rockface with cascading waterfall. Beside it sits the green-and-white-canopied Winter Garden Bistro — which is serving Dutch dishes.

The menu ranges from Soep to . . . well, Gebakken mulfilet in tomatenbouillen (red snapper fillet in tomato Balinese broth). If that sounds like a mouthful, you can order Dadar Bali (Indonesian omelette) or Aardappelen met Goudse Kass (potato crumpets and melted gouda) instead. The maître d' says all 90 seats are filled every evening, "until the horse show starts."

The flower show was designed by Toronto landscape architect Tom Sparling, assisted by Nicole Eaton and Hilary Weston (of Eaton's chain and Weston bread fame). They're the women who co-authored the bestseller *In a Canadian Garden*. The maître d' points to a linen-covered table in front, where a stylish group is tucking into Bahmi Goreng. "In fact," he says, "that's Mrs. Eaton there." And so she is.

In a nearby corner, a man in a black Dutch cap is stooped over a sawhorse, whittling a wooden clog with a poleknife. He is Koos Vreeswijk from Zoeterwoude, Holland, where his family has carved klompens for five generations. And the Netherlands Tourist Board has flown him over to demonstrate his craft.

"I'm fairly new at this," smiles Koos. "The Dutch have been making wooden shoes for five centuries. I've only been doing it for 43 years."

Yet he's happy to show how it's done. First, he says, you saw off a 15-inch piece from a willow or poplar tree, then split it into five sections with a wedge. Each section is then carved with a knife into the rough outline of a shoe. Then the poleknife — which looks like a long straight razor, and is attached at one end to the sawhorse — is used to refine the final shape.

The clogs are then braced, toes down, to the side of the sawhorse, gouged out by hand-turned reamers and scooping chisels, then trimmed with heelhooks and curl knives. Finally, tiny holes are drilled in one side to hang them out to dry before they're painted. And, *voilà* . . . a pair of klompen.

Standing ankle-deep in shavings, Vreeswijk says all shoes were carved by hand in Holland until the mid-twenties, when factories took over. "Since it takes two hours to make one pair," he explains, "there are only a few custom shoemakers still around." Yet, while most wooden clogs today are sold as souvenirs, Koos says, "a million pairs are still sold as footwear. They're not only light and comfortable, but keep the feet dry in our wet polderland."

Then, lesson ended, he turns around and cranks an old Dutch barrel organ called the Windmill. As the organ belts out "Tulips from Amsterdam," Koos Vreeswijk, one of the last great klompenmakers, grins broadly and clacks his poplar clogs.

Beneath a white tent a few yards away sits the Wine Garden. Its carpet and

75 linen tablecloths are green. A massive tree trunk strung with sparkling white Christmas lights rises in the middle — ringed by an awesome selection of sinful desserts. Paper Chinese lanterns hang overhead.

The hostess says the café specializes in fondues. And the wines that give the garden its name? "Ah, yes," she enthuses, "they're the best from Ontario, California and France — as well as those we import."

Huh? Since when did France join NAFTA? We're about to ask if she means "others" rather than "those" — but she's spun off to greet a party of six.

Between the Winter Garden and Coliseum runs the narrow Knob Hill Lane. Stalls along one side display water filtration systems and pumps, and anyone passing can guzzle all the free H_20 they want. On the other side, prize-winning beef carcasses dangle from hooks in a huge glassed-in meat cooler stretching along the wall.

Strange lane! A cross between pump house and abattoir. But a haven for thirsty butchers.

On the opposite side of the Coliseum, the glassed-in cooler of Dairy Lane is far more fun. Surrounded by award-winning displays of Canadian-made cheese (from Havarti and cheddar to Colby and Camembert), two shivering students from the Ontario College of Art crouch inside — carving lifesized sculptures out of giant 50-pound blocks of butter.

In their medium, they're better than Michelangelo — since Signor Buonarroti, brilliant as he was, never once chiselled a single pat. But then, the Renaissance was hardly famed for its fridges.

Through a door off Dairy Lane, you enter the West Annex. On its second floor, the throb of country music resounds from the Western Corral — a big room with a bare wooden floor, jammed with round tables and constant crowds. It's a colourful congregation.

Young parents in sweaters with kids munching burgers. Ma and Pa Kettle in matching caps inscribed, "Satan Enslaves, Jesus Saves." Good ol' buddies in jeans and cowboy boots draining pitchers of overpriced beer. A Dolly Parton blonde in a Three Little Pigs T-shirt. A buxom brunette in a purple-sequined blouse.

Along the walls, short lineups form at kiosks selling draft lager, hamburgers and steaks on paper plates. And everyone seems happy. For up on a stage with a painted Rocky Mountain backdrop, Carl Kees and his Golden Fiddle Music Co., a quartet topped by feathered Stetsons, are twanging out Western oldies. And shoes are tapping in rhythm across the room.

Between songs, Carl touts tapes of his group's recordings. "I really think you should buy a couple, folks," he says. "They're collectors' items. The last time we made one, they kept the tape — and released us."

Even Ma and Pa Kettle chuckle.

Behind the Western Corral is the Christmas Village, a maze of shops and display booths selling everything from stirrups and saddlesoap to turquoise rings and scented candles. Strolling the aisles to the rear of the floor, there's a long, snaking lineup of children and adults. A huge banner hanging above them reads: Big Ben's Retirement Tour.

And there in a paddock, draped in a blue satin cooler and held by a groom, stands the great chestnut champion himself — Canada's most celebrated show jumper ever. For a nation that reveres its athletic stars, the 18-year-old gelding is among the most loved . . . and honoured. In a career spanning 11 seasons, Big Ben won more than 40 grand prix, performed for three Olympic teams and became the only North American show jumper to earn more than $1.5 million.

Brought to Canada from Holland in 1983 by his owner, the rider Ian Millar, Big Ben became famous not only for his will to win but also, after three close brushes with death, to live.

Now, after a triumphal tour across Canada, he's making his last official appearance before retirement at the Royal. "From here," says an attendant, "Big Ben will go home and gallop free." Beside the queue stretched out to bid goodbye (more than 3,000 on this final day alone), there's another line of fans who've paid $7 each (split between Canada's Equestrian Team and its Therapeutic Riding Association) to pose for photos with the solemn star.

The young blonde photographer snapping pictures keeps smiling, but looks bushed. "Most days we take about 300 shots," she says, "but last Saturday it was more than 700. And today, it's just gone crazy." She rolls her eyes. "It's like shooting closeups on a roller-coaster."

To the west of the annex, a corridor connects to the Royal's real realm of bluebloods — the massive two-storey Horse Palace.

In one corner of the second floor sits the bustling Fox & Hound restaurant, with its rustic touches of hunting scenes, stained-glass windows and saddles displayed on columns. At the tiny bar every stool is taken, and across the room there's a lineup at the low-priced take-away counter. Chicken breast sandwich: $3.95. The daily special, Polish Delight: $4.99.

"We're always busy," says the manager, who's sitting at the bar. "We get up to 200 patrons in here daily. But then, we're *always* open. It's the only restaurant running 24 hours. Daytimes we pull in the middle-of-the-road crowd. Nights we get the exhibitors and stablehands. Even at five in the morning, they can get a hot sandwich."

He points to a small raised stage in a corner. "We also have live entertainment every night." What kinds of groups perform? "Well," he smiles, "none I'd wanta pay for. That's our karaoke stand." There are chuckles at the bar.

A large window in the Fox & Hound faces directly onto a white-fenced show ring called Cadillac International Stallion Lane. Here, three times daily, owners and breeders display their prized Arabians, thoroughbreds, heavy horse breeds and Canadian Sport Horses in the popular "Parade of Stallions."

A crowd is clustered around the wooden rail as a stunning chestnut named State Affair is slowly circled around the sawdust by a groom.

There's a burst of applause when a woman in a broad-brimmed black hat and scarlet jacket announces that the horse, owned by Big Curve Acres north of Barrie, "is the son of the great Secretariat.

"Unfortunately," she says, "State Affair — a spitting image of his sire — didn't have the same chance to prove himself racing, since he damaged his ankles as a two-year-old." But he's sound for recreational riding, she adds, and with stud fees there's a live foal guarantee.

And how much, we ask when the show ends, is State Affair's stud fee? "That's something to discuss with the owner over there, Henry Regelink," she says pointing. "But I think you'll find it's a fraction of his father's."

Which was? She smiles sweetly. "A cool hundred thousand."

Just past Stallion Lane, red and white bunting and bright-coloured streamers dangle below the glass skylight. Here another standing crowd elbows around a railing, staring down at a walking ring banked by bleachers on the floor below. A dozen young stablehands in jeans and jodphurs are leading or riding their horses around the oval.

Yet most eyes are on Doug Henry of Canada's equestrian team, who canters his steed around the boards — like a star warming up in a group of extras. But Doug is as nonchalant as the others. They're all just out there exercising horses, circling the ring, hoofbeats smothered in the earth. It's like watching a silent movie. After a few restful moments, most spectators drift away. Because, all around, the myriad sounds of the Horse Palace echo like muted thunder. The sounds are the clatter of horseshoes on cement, the rumble of wagon wheels, the sharp bark of dogs, the snorts of stallions, the overlapping waves of conversation and sudden gusts of laughter.

Between broad concrete aisles, rows of adjoining high-barred stalls stretch the length of the massive room. Some are open, with sleek horses staring back at the passing parade. Some are shrouded with canvas tarpaulins, others draped with velvet.

Small dogs are tethered in front of many stalls. Outside others, stablehands with watering cans dampen down the dust. Some are proudly pinning prize ribbons onto corkboards. Some sit on stools polishing harness. Others stand on wooden crates braiding manes. "The mares get twenty-one braids, the stallions twenty-two," says one busy lass. "It takes about an hour."

She grins down from her box. "Why do you ask? You want me to do you next?"

Down a flight of stairs, the vast main floor of the Horse Palace is reserved for the massive draft horses. Here too, large blocks of stalls with personal tack rooms are held by wealthy owners. One Schomberg stable alone has 20 horses at the fair — with nearly 50 prize ribbons already on display. Famous farms from such northern states as New Jersey and Pennsylvania also display their ornate regalia.

Five draped stalls hold horses owned by an Illinois stable. A young lady in monogrammed blazer sits in front of its tiny tack room. Inside, it's like a tall tent with straw-strewn floor — as designed by MGM. An Ivanhoe set before a joust. A Crusader knight's bivouac during a siege.

Rich maroon drapings rimmed with gold trim line the walls — which are hung with six sets of gleaming black leather harness. A lighted chrome chandelier dangles overhead. Atop a tiny table sits a silver mint-filled bowl. It's the perfect nook for a champagne tryst. The only thing lacking is Sir Galahad.

The young lady strides across the alley outside (on which golden horseshoes are stencilled as guidelines) to a cluster of gleaming four-wheeled wagons and carriages parked side by side. "This," she says, pointing at one, "is our Park Drag coach, built in 1902. They were only bought," she smiles, "by the very rich. I've no idea what it cost initially. But you'd pay anywhere from fifty to a hundred thousand today."

"The coach is pulled," she says, "by a four-in-hand. The two horses in front are the 'leaders.' The two in back are the 'wheelers.' And the driver's called the 'whip.'"

We say that the inside seating space seems confined. "Yes, cramped, claustrophobic *and* dusty," she laughs. "The only comfortable seat is the driver's."

Farther down the lane a small crowd is gathered in front of eight stalls. Inside are the world-famous Budweiser Clydesdales.

"Well, in fact," says a groom, who's shining harness, "these are just *some* of them. Actually, the brewery raises nearly three hundred Clydesdales on ten different farms. It also has three separate teams performing at Sea Worlds in Florida, Texas and California. This is just one of the three *travelling* teams, which operate out of Missouri, New Hampshire and California."

He says they cover more than 90,000 miles each year, and appear in over 300 events. Each team travels in a three-van caravan with seven handlers: the hitch driver, assistant driver and the chauffeur grooms.

Two of the vans (custom-built with thick rubber flooring, air suspension and vent fans) carry the horses. The third holds the brewery wagon, portable stalls, harness, trappings and special feed — "a mixture," he explains, "of crimped oats, beet pump, brans, minerals, salt, molasses and water."

Sounds delicious. "Well," he says, grinning, "it's healthy! Each horse eats up to twenty-five quarts of feed a day — as *well* as more than fifty pounds of hay and thirty gallons of water. But they're *big*, don't forget. Eighteen hands at the shoulders. And the biggest weigh 2,300 pounds. More than a ton." Yet only the finest Clydesdales make the show teams, he says.

The ideal horse is bay in colour, has a black mane and tail, a blaze of white on the face, and white feathering on all legs and feet. Still, no matter how perfect, it takes five hours to groom the horses before a performance. They are shampooed and brushed, their feathering fluffed white, their manes and tails braided and woven with red and white ribbons and fresh flowers. Then — after the silver and brass on their harness (worth $50,000) has been polished — it takes another 45 minutes just to hitch them.

The first two horses, called the "wheel pair," are positioned next to the classic brewery wagon (which alone weighs four tons), followed in order by the body pair, swing pair and lead pair. Finally, the uniformed "hitch" and his assistant climb atop, along with the traditional Dalmatian mascot. "The drivers," says the groom, "are amazing guys. They guide those seventy-five-foot, eight-horse hitches with forty pounds of reins in their hands."

As for the Clydesdales themselves, they were first bred on the banks of Scotland's River Clyde, brought to Canada in 1842 and used for heavy transportation until trucks (the new "horsepower") took over. Fortunately, in 1933, August Busch Jr. gave his dad a team of Budweiser Clydes, plus a bright red Studebaker wagon, to celebrate the repeal of Prohibition. Today, they're the symbol of the world's largest brewery.

Inside one stall, three small blonde girls clutching Barbie dolls are curled up on blankets on top of the straw.

Outside, an elderly groom in a stained green fedora is brushing a huge black horse. A father and small son stand watching.

"Man, he's a beauty," says the father.

"You bet," agrees the groom, beaming. "But they *all* are." He points his brush around the room. "If you wanna see what makes a great horse great, this here's the place to be."

The father lifts his son to stroke the stallion's flank. "I *love* horsies," the boy murmurs shyly.

"Me too, son," the old man says, smiling. "Me too."

3 *A Night at the Show Ring*

"THERE ARE, GOES A SMALL WISDOM FREQUENTLY repeated around the horse-show circuit, two things a rider definitely needs. Money and more money. And more money.

"Anyone who has ever been involved in the horse world learns very quickly to appreciate that daunting reality. Showing horses is a costly pastime.

"A talented jumper — a horse game and agile enough to take on a course filled with obstacles and jumps — will cost perhaps $30,000. But a great horse, one that's strong, smart, fast and beautiful — a 1,500-pound animal that can be ridden by a world-class athlete against a ticking clock — will go for more than $1 million. Worse, buying the horse is simply the entrance fee. Barn and board can run to $2,000. A month. Then there are trainers and vet bills to pay. And tack to buy. Shoeing can cost $200 a month. A Hermes saddle can be as expensive as $6,000. Vogel boots, custom-fitted, can run $1,000 a pair. And if you're going to do the [weekend] circuit, your horse is going to have to be shipped, unless you want to buy a van. It is a very expensive sport.

"But show riding is more than a sport. Being 'horsey' is a way of life. For some, it is an aristocratic heritage, an almost atavistic instinct passed on with the family name. For others, it is an acquired, but no less defining, ritual. It is an activity largely practiced by a certain class — and those who aspire to belong. It is, on its carefully preened, vaguely English surface, a pursuit practiced in a comfortable, clubby world of white-fenced paddocks, rolling lawns, tailgate parties and Jack Russell terriers, of tweedy gentlemen and ladies who wear pearls. Ralph Lauren has made a fortune by shrewdly mimicking its habits and pretentions."

Howard Blum, *Vanity Fair*, January 1995

Most fairgoers don't even know it exists. But in a second-floor corner of the East Annex, sprawl the posh salons of the Tanbark Club. Named for the leather scraps discarded by tanneries, and once used as footing in the Coliseum, it was founded in 1922 when the Royal first began.

The club today has 350 members, most with very Anglo-Saxon surnames, who hold private sets of box seats for the horse show. It's been called the most expensive club in Toronto, because most of its members congregate each year for just 12 days — the duration of the fair.

But during that time, the Tanbark members, in their tuxedos and gowns, dine and drink with the horse set, the riding teams and Royalty — since the fair tries to get some member of the British monarchy to open the show each year. Thus such horse-loving princesses as Anne, Margaret and Alexandra, as well as the Duchess of York and Prince

Philip, have all done the honours. The "President's Room" is the club's inner sanctum, with soft green carpet, marble-papered pillars, brass chandelier and avocado-coloured stucco walls. Sofas, small tables, soft easy chairs, flower-filled vases and lamps handpainted with horses are arrayed throughout, divided by silk screens with ornate fruit designs. A table is laden with delicate hors d'oeuvres. A bar stands in one corner.

The walls are hung with oval gold-leaf mirrors, framed prints of top-hatted riders surrounded by hounds, carriages on the road to Epsom Downs and a portrait of Lady Eaton astride her horse, Paddy.

A doorway leads into the "Directors' Boardroom," with green chairs against the walls. Besides the potted ferns, ceiling fans and corner bar are photos of the club's ex-presidents, superb watercolour paintings of prairie grain elevators and three large trophy cases filled with silver cups.

Finally, there's the spacious "Bird Room" near the entrance, again with the ubiquitous trophy cases displaying giant "Challenge Cups" — named for the dukes of both Bedford and Westminster — as well as an actual Canadian trophy: the 1945 Governor General's Cup. Small tables are scattered throughout the Bird Room, and a buffet is set up beside the bar.

Before the afternoon and evening horse shows, members flow into all three Tanbark suites, where attentive servers in red jackets await. Most go for cocktails and insider chatter, then drift off to dine in one of the restaurants.

At night many members head to the posh Café Royale, a few steps away in the upstairs East Annex, the fair's most formal, and expensive, dining area. At the café's entrance, a fountain spills behind a decorative setting of driftwood and live pines. A headwaiter checks for the required reservations.

On one side of the spacious room a tuxedoed pianist at a baby grand plays soft Gershwin tunes, while under muted chandeliers, tables are topped with white linen and candles. Most guests are bedecked in formal dress, and all 220 seats are taken. A buzz of voices mingles with the music. It's an atmosphere of relaxed elegance.

The "Gala Buffet," served between 5:30 and 8:00 p.m., stretches along one wall. It's a feast of patés, lobster, chicken, fish, beef and a rainbow of rich desserts, and costs a cool $42.50 — *before* the vintage wines are even chosen.

Still, by the time the horse show starts, many diners have departed for the night's extravaganza.

The vast Coliseum, with its tiers of boxes and bleachers soaring high on all sides, is the central building of the fair — in terms of both its location and its series of spectacles. But the horse shows are only the highlights. During the Royal's twelve-day run, the Coliseum pulses with action from daybreak to midnight.

As early as 7:00 a.m., a variety of dairy and beef cattle (from Holsteins to Aberdeen Angus) are paraded around the huge arena to be judged. Some mornings, when the city is still waking, the ring will be filled with Belgian four-horse teams or Hunters hurtling fences. Even the Canadian Equestrian Team often circles the sand at breakfast time.

And so it goes throughout the day. Open Jump competitions. Hackney Ponies. Jersey and Hereford judging. Clydesdale Two-Horse Teams. Percheron Six-Horse Teams. Even the Superdogs. As kiosks on the upper level hustle hotdogs, pop and pretzels, the Coliseum shows switch rapidly. Some families camp in the stands for hours. Others go only for specific events.

But most top performances take place after dark. And every night is different. They're the shows most Tanbark members have come for. By the time the first riders enter the ring, most boxes are filled with the glittering elite. Yet the entire Coliseum is also packed.

It's a fascinating crowd. City slickers, farmers. Old couples, dating youngsters. Rich breeders, stablehands. It's a mixture of tuxedos and T-shirts, of top hats and Stetsons — the Old Guard versus the Old West.

In the arena, the "Combined Working Hunter Stake" is under way. Despite its title — which, like those of most Coliseum events, sounds as exciting as a cement-mixing contest — the show is fun.

Twelve riders in traditional black helmets, scarlet jackets, white breeches and gleaming black boots are jumping sleek steeds over fences. Yet, with all nine fences painted green and white, and festooned with large cut-out shamrocks, it looks like some Dublin Derby.

A portly photographer in a bulging tux lurks behind two pine trees planted near the last fence. As the riders make their final leap, he pops out to take shots, then whips back behind the branches.

Since the hunters' hoofbeats are muffled in the sand, the only sound is the murmur, or applause, of the crowd. Only when the riders gallop to the exit gate do you hear the slap of their gloves on the horses' necks.

The instant the last rider disappears, a huge gate on the opposite side of the arena swings open — and a small invading army, wearing blue caps and coveralls, races in. It's led by four shiny blue tractors towing flatbeds, onto which the thirty-man crew swiftly pile the fences, pine trees and potted geraniums. Like a team of crack commandos, they're in, stacked up and out in under three minutes.

As the last flatbed exits, a fifth blue tractor dragging a mesh net roars in and starts smoothing the sand in diminishing circles. The driver speeds so fast, in fact, that his huge rear tires skid sideways during the wild, careering arcs. It's a show in itself, a one-man tractor Grand Prix. When the ring's been swept flat and he aims for the exit, the delighted crowd applauds.

The tractor dust is still settling when, from a bandstand high in the north-end bleachers, comes a blast of brass horns. Echoes of Madrid on a Sunday afternoon. You almost expect a snorting toro to lunge through a gate as El Cordobes strides stiff-legged towards him.

Instead, from one side of the ring, two uniformed cadets roll a long red carpet onto the sand. Their gangly comrades place a table on the carpet and pile it with trophies. Then, to another fanfare, three champion cows — an Ayrshire, Jersey and Brown Swiss — are led in, and trundled across the ring to receive their cups and ribbons.

Moments later the arena is awhirl with seven separate four-horse teams pulling painted wagons. Racing around the boards in their brightly braided manes, plumes and sparkling harness, the teams of Belgians, Percherons and Clydesdales are magnificent. It's a spectacle of total glitz and dazzle.

When the teams have raced out, there's another blast from the bandstand, and the Governor General's Horse Guards march smartly into the ring. It's a proud and famous regiment, having served its country from 1912 and the Crimea through both World Wars and Korea.

Behind them, Ontario's Lieutenant Governor Henry (Hal) Jackman, resplendant in medals, rides into the arena in a pitch-black caleche, seated behind a top-hatted driver who is madly chewing gum. As a brass band thunders, Hal inspects the troops, takes the march past, and presents (to a steed named Signature) the silver Lieutenant Governor's Cup.

When the madman on the tractor has seethed again around the sand, drivers in jog carts sweep in behind high-stepping road horses. Once the trotters are walked, then jogged, around the ring for the judges' appraisal, there's a shouted command — and the horses shift into high. They race their fragile carts so swiftly on the curves that the drivers lean far over the sides like sailors on listing yachts.

In the photographers' pen on the west of the ring, cameras click like crazed cicadas.

For the next event, the Puissance for International Teams and Open Jumpers, tuxedo-clad TV cameramen circle the arena. This competition, with total prizes of $15,000, is one of the Royal's top attractions. In it, six scarlet-jacketed riders attempt to hurdle a wall in three separate jump-offs — in which the wall is consistently raised.

Still, no matter the riders' expertise, the steeds are the stars. For

"Puissance" means power or strength. And the jumpers have it. Legs and muscles like steel suspension springs. They canter effortlessly up to the walls — then arch in the air as if shot off a trampoline. Up and over the six-foot structures.

On a few jumps, the horses' rear hoofs dislodge the top block of the wall, for lost points. In one of the final jump-offs, a horse approaches a wall and suddenly balks, swerving instead to the left. He's scratched. After each round of jumps, men in coveralls dash out and rake the sand, while TV announcers lean across the boards to interview the riders.

In the second jump-off, the walls are raised as high as six foot ten. Although more blocks are toppled, many of the jumps appear effortless. With each successful hurdle, the crowd cheers. One of the equestrians drawing loud ovations is Ian Millar's son, Jonathon, who is riding two separate horses in the Puissance.

In a brightly lit booth set high in the eastern bleachers, two rows of judges, timers and announcers sit at tables. They crane forward during the third and final jump-off as the gate is raised to seven foot one. Finally, only three jumpers are left in the competition. But instead of pushing for the wall to be raised, they decide their horses have reached their limit. So they share the title and split the prize money.

The Coliseum echoes with cheers as they exit the arena.

The shouts ring out again as the eight Budweiser Clydesdales with bobbing white plumes thunder in, the red wagon stacked high with beer cases and the Dalmatian grinning on the seat beside the drivers.

The orchestra then blares for the start of the "Road Horse Stake." Eleven elegant horses hauling road wagons circle the arena, their poker-faced drivers in charcoal suits and narrow-brimmed fedoras. The entrants are judged on jog trot, road gait and speed, and the winning owner (who looks as delighted as a man with mumps) is declared the Royal and Canadian champion.

Next in the ring is the "Six-Horse Team Championship Series" — three teams of Belgians, Clydesdales and Percherons in silver harness, pulling sparkling red wagons with lit side lamps. But even more Dickensian is the judge, who solemnly scuttles from team to team, a round-shouldered beachball in tight-fitting tux, with the weirdest chapeau on his head: a *short* top hat.

Four more six-horse teams hit the ring, manes festooned like Christmas trees, as the orchestra belts out "The Yellow Rose of Texas." The rigid drivers, clutching buggy whips, sit solemn as Amish elders. The galloping teams thrash around the arena — a wave of huge bobbing heads, pounding hoofs, flashing silver. The director John Ford would love it. It's the Oklahoma Land Grab in Cinemascope.

Minutes later, when six coaches charge in, Dickens truly enters the arena. Each one is pulled by four geldings and packed with a dozen elegant passengers — including gents who stand at the back blowing into long brass coach horns. It's a scene from the *Pickwick Papers*. The show is called the "Four-in-Hand Suitable for Pleasure." Another inane title — but pleasure it *does* provide.

In the press box, a veteran horse show writer from New Jersey explains the event. "The top prize, you notice, is a paltry $150. Since the coaches cost up to $100,000, each horse about $50,000 and harnesses

$10,000 or more, the entrants, obviously, couldn't care less about the cash. It's strictly a *very* rich man's show. Yet the same people compete around the world. They just love to get together and party every night.

"But those guys with the coach horns are pros. One of them plays with the Philly Philharmonic. And they're blowing authentic century-old calls. Each one means something. Like, 'Watch out, we're turning,' or, 'It's foggy ahead.' In the early days, the gentry crowded on top like they're doing here, and the servants sat inside."

She smiles. Wryly. "Today the servants wait in the tack rooms chilling champagne."

The tractors and workmen race back, and swiftly erect the fifteen fences for the Grand Prix Challenge. Then 21 riders in red or black jackets gallop in — members of the Canadian, Swiss, German and American equestrian teams. Out of the four major indoor shows each year, in Pennsylvania, Washington, D.C., and New Jersey, the one at the Royal Winter Fair is the last. And prizes total $30,000 — ranging from $10,000 for first to $200 for tenth.

The riders have 81 seconds to make all 15 jumps, and a huge digital clock above the arena clicks them off. The riders are judged by time and fences cleared. At the end of the first round, only the eight top competitors with the fewest faults and ties are allowed to return for the jump-off.

As the Challenge continues, fences topple, horses balk, faults increase. But scores of jumps are sensational. Ian Millar, riding Mistral, receives constant applause. In the final jump-up there are only nine fences for the eight remaining horses, and the time limit's down to 52 seconds. When it's over, the Canadians have come in second, third and fourth. An American has won.

The band plays "Yankee Doodle Dandy."

The audience erupts as the maniac on the tractor blasts in for the final time — resplendent in tuxedo. And the work crew erects a huge, flower-laden horseshoe in the centre of the ring.

The Coliseum hushes as the overhead lights dim. And the deep booming gongs of London's Big Ben fill the room. Then, centred in a spotlight, Ian Millar gallops through the gate on the great clock's famous namesake.

The applause is raucous as the arena explodes with a barrage of flashbulbs — like heavy flak in a night sky. Big Ben slowly circles the boards, as a beaming Millar bends over to touch scores of outstretched hands. After their third triumphal circuit, horse and rider race down the middle of the ring and leap into the horseshoe. It's pure Flo Ziegfeld-style showmanship.

Then Big Ben's small blonde groom, Sandi Patterson, and Millar's wife arrive in a carriage to join them. Cadets in red tunics and white pith helmets march in to stand as honour guard. Inside the horseshoe, the huge chestnut snorts as wagons roll in bearing farewell gifts — bushels of apples, oats, carrots and bran muffins.

Then, suddenly, the gelding starts kicking his hoofs. Bucking wildly, he sends pots of flowers flying in the air as tiny Sandi swings from his reins. The audience is delighted.

When Ben is finally calmed, the Canadian equestrian team gallop in and raise their helmets as they ride past the champ. And finally, a Winter Fair official drapes a gold-trimmed blue collar on the horse. It reads: "Big Ben. Thanks from the Royal." Ian Millar takes the mike to salute his "ton and a half of dancing dynamite." Ben, he says, "is the bravest, best, most intelligent horse I've ever worked with. There's no better example of excellence. Tonight is not a retirement, but a celebration. Tomorrow, we'll go for a ride and talk about things."

The crowd goes wild — until Millar asks everyone to rise and hold hands. Then he prays. "Oh Lord, we start a new phase of life now, and may your mercy be long." Then, swinging back up on Ben, the two champions again slowly circle the ring, Millar touching every outstretched hand. The Coliseum is still thundering as the band booms into "O Canada."

And so the Royal ends for another record-setting year.

Hundreds of fairgoers pass through the Winter Garden into the night. Most are clutching pots of red and yellow tulips.

With many thanks to Barbara Fulton, for her enthusiastic assistance in researching this book during long (but daffily delightful) days and nights at the fair.

To Renée Dykeman, for her dexterous editing.

And the staff of the Royal, for steering us through the teeming 27-acre labyrinth and so generously allowing access to seven decades of archival material.

PAUL KING